Push your Career Publish your Thesis

Science should be accessible to everybody. Share the knowledge, the ideas, and the passion about your research. Give your part of the infinite amount of scientific research possibilities a finite frame.

Publish your examination paper, diploma thesis, bachelor thesis, master thesis, dissertation, or habilitation treatises in form of a book.

A finite frame by infinite science.

An Imprint of
Infinite Science GmbH
MFC 1 | Technikzentrum Lübeck
BioMedTec Wissenschaftscampus
Maria-Goeppert-Straße 1
23562 Lübeck
book@infinite-science.de
www.infinite-science.de

Editor

Thorsten M. Buzug
Institute of Medical Engineering
University of Lübeck
buzug@imt.uni-luebeck.de

Reihe: Medizinische Ingenieurwissenschaft und Biomedizintechnik

Diese Reihe umfasst Werke der Medizinischen Ingenieurwissenschaft und Biomedizintechnik, deren Themen strategisch unter den Zukunftstechnologien mit hohem Innovationspotenzial anzusiedeln sind. Als wesentliche Trends dieser Forschungsgebiete, sind die Schlüsselbereiche Computerisierung, Miniaturisierung und Molekularisierung zu nennen. Bei der Computerisierung sind dabei die inhaltlichen Schwerpunkte beispielsweise in der Bildgebung und Bildverarbeitung gegeben. Die Miniaturisierung spielt unter anderem bei intelligenten Implantaten, der minimalinvasiven Chirurgie aber auch bei der Entwicklung von neuen nanostrukturierten Materialien eine wichtige Rolle, und die Molekularisierung ist in der regenerativen Medizin aber auch im Rahmen der sogenannten molekularen Bildgebung ein entscheidender Aspekt. Forschungs- und Entwicklungspotenzial werden auch der Biophotonik und der minimal-invasiven Chirurgie unter Berücksichtigung der Robotik und Navigation zugeschrieben. Querschnittstechnologien wie die Mikrosystemtechnik, optische Technologien, Softwaresysteme und Wissenstechnologien sind dabei von hohem Interesse.

Matthias Weber

Power-Loss Optimized Field-Free Line Generation for Magnetic Particle Imaging

Medical Engineering Science and Biomedical Engineering — Volume 4

Editor: Thorsten M. Buzug

© 2015 Infinite Science Publishing
the BioMedTec Science Campus Publisher Lübeck

An Imprint of Infinite Science GmbH,
MFC 1 | BioMedTec Wissenschaftscampus
Maria-Goeppert-Straße 1
23562 Lübeck

Cover Design, Illustration: Uli Schmidts, metonym
Copy Editing: University of Lübeck, Institute of Medical Engineering

Publisher: Infinite Science GmbH, Lübeck, www.infinite-science.de
Print: Books on Demand GmbH, Norderstedt

ISBN Paperback: 978-3-945954-06-5

Das Werk, einschließlich seiner Teile, ist urheberrechtlich geschützt. Jede Verwertung ist ohne Zustimmung des Verlages und des Autors unzulässig. Dies gilt insbesondere für die elektronische oder sonstige Vervielfältigung, Bearbeitung, Übersetzung, Mikroverfilmung, Verbreitung und öffentliche Zugänglichmachung sowie die Einspeicherung und Verarbeitung in elektronischen Systemen.

Die Wiedergabe von Gebrauchsnamen, Handelsnamen, Warenbezeichnungen usw. in dieser Publikation berechtigt auch ohne besondere Kennzeichnung nicht zu der Annahme, dass solche Namen im Sinne der Warenzeichen- und Markenschutz-Gesetzgebung als frei zu betrachten wären und daher von jedermann verwendet werden dürften.

Bibliografische Information der Deutschen Nationalbibliothek:
Die Deutsche Nationalbibliothek verzeichnet diese Publikation in der Deutschen Nationalbibliografie; detaillierte bibliografische Daten sind im Internet über http://dnb.d-nb.de abrufbar.

Bibliographic information published by the Deutsche Nationalbibliothek
The Deutsche Nationalbibliothek lists this publication in the Deutsche Nationalbibliografie; detailed bibliographic data are available in the internet at http://dnb.d-nb.de.

Contents

1. **Introduction** — 1
2. **Fundamentals** — 5
   - 2.1 Electric Current — 6
     - 2.1.1 Current I — 6
     - 2.1.2 Current Density $\mathbf{j}$ — 6
     - 2.1.3 Continuity Equation — 7
     - 2.1.4 Ohm's Law — 7
     - 2.1.5 Electrical Power — 7
   - 2.2 Fundamentals of Magnetostatics — 8
     - 2.2.1 Biot–Savart Law — 8
     - 2.2.2 Maxwell's Equations — 9
     - 2.2.3 Magnetic Moment — 9
     - 2.2.4 Magnetostatics in Matter — 10
   - 2.3 Electrodynamics — 14
     - 2.3.1 Faraday's Law — 15
     - 2.3.2 Maxwell's Extension — 15
   - 2.4 Maxwell's Equations — 15
   - 2.5 Hall Probe — 16
     - 2.5.1 Hall Effect — 16
   - 2.6 Magnetic Particle Imaging — 18
     - 2.6.1 Signal Generation — 20
     - 2.6.2 Resolution — 24
     - 2.6.3 Reconstruction — 26
3. **Materials and Methods** — 31
   - 3.1 Introduction and Coil Configuration — 32
   - 3.2 Coil Controlling — 34
   - 3.3 Coil Construction — 35
     - 3.3.1 Coil Forms — 35
     - 3.3.2 Litz Wire — 36
     - 3.3.3 Winding Process — 38
   - 3.4 Coil Assembling — 39
   - 3.5 Scanner Case Design — 42
     - 3.5.1 Permanent Magnet Plates — 43
     - 3.5.2 Air Cooling System — 44

Contents

		3.5.3	Coil Configuration Sealing Part	45
		3.5.4	Air Outlet System	45
		3.5.5	Scanner Assembling	45
	3.6	Magnetic Field Measurements		45
		3.6.1	Measurement Environment	46
		3.6.2	Parameter Setup	47

4 Results — 51
- 4.1 Coil Assembling — 52
- 4.2 Field Measurements — 54
  - 4.2.1 Permanent Magnet Adjustment — 54
  - 4.2.2 Results for FFL Rotation — 55
  - 4.2.3 Results for FFL Translation — 56
  - 4.2.4 Power Loss — 58
- 4.3 Scanner Case — 59
  - 4.3.1 Air Cooling System — 59
  - 4.3.2 Plate Magnet — 60
  - 4.3.3 Assembling — 62

5 Discussion — 63

6 Conclusion — 67

7 Bibliography — I

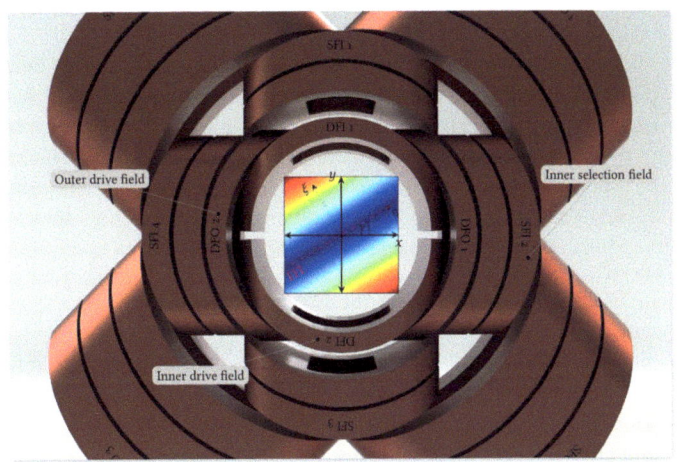

Chapter 1

Introduction

1 Introduction

The new imaging modality Magnetic Particle Imaging (MPI) is capable of visualizing the distribution of superparamagnetic iron oxide nanoparticles achieving high spatial and temporal resolution. This method was first published by Gleich et al. in 2005 [1]. In 2009, first in-vivo images of a beating mouse heart demonstrated the high potential [2].

MPI uses magnetic field configurations generating a gradient field that possesses a field free point (FFP). This FFP is moved through the investigated area. A characteristic signal can be measured in the receive coil. Afterwards, the signal is reconstructed to the spatial distribution of the particle concentration. Clinical applications of this quantitative imaging process could be especially scenarios where a fast and dynamic imaging is necessary. Diagnostics of coronary vessels and possible vasoconstrictions respectively angiography could be possible [3]. Here, no ionizing radiation or harmful tracers are used.

In 2008, an enhancement of the FFP scheme was introduced using a field free line (FFL) to facilitate spatial encoding [4]. Simulation studies showed a sensitivity gain of one order of magnitude. This still allows signal generation at low particle concentrations when signal-to-noise ratio (SNR) decreases at conventional FFP scans. However, an implementation of this concept was not feasible since 16 coil pairs were used to generate the gradient field resulting in a thousand times higher power consumption compared to an FFP scanner with same gradient strength and equal size [5]. Furthermore, an expensive and complex cooling system would have to be implemented. Nevertheless, further simulation studies could reduce the amount of coils and consequently reduce the power loss. The electrical power consumption was reduced to values comparable to those of FFP scanners and maximal four coil pairs were needed. In 2011, the first FFL generation proofed the concept of magnetic field generation for dynamic FFL imaging [6]. A gradient of $0.25\,\text{T}\,\text{m}^{-1}\,\mu_0^{-1}$ was realized.

A customized reconstruction technique for FFL imaging demonstrated feasibility for fast image reconstruction [7]. No time-consuming system matrix has to be measured and no memory intensive matrix inversion algorithms have to be applied. Premise for functionality is high field homogeneity along the FFL as well as parallel to its alignment.

The newest optimization proposes curved rectangular coils as a requirement for efficient FFL generation [8]. In comparison to a conventional design with circular coils, the field quality is enhanced by a factor of five and the electrical power consumption by a factor of almost four. The presented work experimentally validates these simulations by constructing customized curved rectangular coils being assembled to a complete FFL selection and drive field coil configuration. Furthermore, a scanner case is designed fixing the construction and building the fundament for a complete scanner setup facilitating dynamic FFL imaging. The main requirement for the case is a customized air cooling element keeping the coil temperature at a minimum. This setup is evaluated by measuring the magnetic fields of the FFL in an appropriate measurement environment. Subsequently, these fields are compared and validated to the simulation studies. Additionally to the field quality, the power consumption is analyzed.

The work in hand is structured in the following way: first, the physical and technical fundamentals of MPI are explained in chapter 2. Here, magnetostatic and electrostatic phenomena describe field generation and interaction. Signal generation, spatial encoding, and reconstruction techniques frame the basis of MPI. Chapter 3 summarizes the construction process of coils, coil assembly, and scanner case. Moreover, measurement environment and the procedure evaluating the generated fields are explained. Subsequently, the results are presented in chapter 4 and discussed in chapter 5. Finally, chapter 6 summarizes the observations and gives an outlook to further work.

Chapter 2

Fundamentals

Physical and technical fundamentals are an important basis to understand, optimize and engineer imaging systems. The imaging modality magnetic particle imaging (MPI) requires elementary knowledge of magnetic fields, their generation respectively interaction, and further mathematical discussions concerning reconstruction techniques. The following sections cover magnetostatic and electrodynamic principles. Furthermore, the functionality of MPI is introduced by describing signal generation and reconstruction techniques afterwards. The physical Fundamentals are mainly summarized from [9].

2 Fundamentals

2.1 Electric Current

Electric current in a conductor generates magnetic fields and is an important parameter defining characteristics of an MPI scanner. Before taking a closer look at magnetic fields, several dimensions by means of the electric current are introduced.

There are two possibilities to realize a sorted movement of electrical current:

1. Either by moving a charged object through space, or
2. by creating a potential difference between both ends of an electrical conductor – applying a force on effectively free charge carrier.

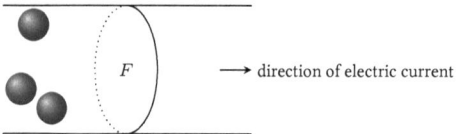

Figure 2.1: Schematic diagram describing the relation between charged particles and the current. The electric current flows from left to right through the cross section F [9, p. 162].

2.1.1 Current I

The current I [A] is the amount of charge dQ [C] that crosses a certain profile F in a specific time (see figure 2.1)

$$I = \frac{dQ}{dt}. \tag{2.1}$$

2.1.2 Current Density j

The current density is a vector **j** with a direction that is collinear with the moving direction of the electrical charge. The absolute value corresponds to the current that interfuses an area per time

$$|j| = \frac{I}{F} = n \cdot q \cdot v. \tag{2.2}$$

In the general case, the current density is a time dependent vector field

$$\mathbf{j}(\mathbf{r},t) = \varrho(\mathbf{r},t) \cdot \mathbf{v}(\mathbf{r},t) \tag{2.3}$$

that is connected with the electrical charge density $\varrho(\mathbf{r},t)$ and the velocity field $\mathbf{v}(\mathbf{r},t)$ of the system.

2.1.3 Continuity Equation

The following relation

$$\frac{\partial \varrho}{\partial t} + \text{div}\,\mathbf{j} = 0 \tag{2.4}$$

can be derived from the Gaussian law (see [9, p. 163]). Here, only the static case $\frac{\partial \varrho}{\partial t} = 0$ is observed and accordingly

$$\text{div}\,\mathbf{j} = 0. \tag{2.5}$$

One consequence is e.g. Kirchoff's current law [10, p. 55 ff.].

2.1.4 Ohm's Law

One can experimentally show that the current I through an electrical conductor is proportional to the voltage V across the ends

$$U = I \cdot R. \tag{2.6}$$

The resistance $R\,[\Omega]$ is temperature dependent. This means that for higher temperatures – e.g. heat evolution caused by an electric current – the resistance rises.

2.1.5 Electrical Power

If the charge q is moved in an electric field $E\,[\text{V}\,\text{m}^{-1}]$ by the distance $d\mathbf{r}$, the work

$$dW = \mathbf{F}(\mathbf{r}) \cdot d\mathbf{r} = q\mathbf{E}(\mathbf{r}) \cdot d\mathbf{r} \tag{2.7}$$

is done. The derivative concerning the time dt

$$\frac{dW}{dt} = q\mathbf{E}(\mathbf{r}) \cdot \mathbf{v}(\mathbf{r}) \tag{2.8}$$

is the electrical power. Now, q can be substituted with the electrical charge density $\varrho(\mathbf{r})$ (see equation (2.2))

$$dP = \left[\varrho(\mathbf{r})d^3r\right]\mathbf{E}(\mathbf{r}) \cdot \mathbf{v}(\mathbf{r}) = \mathbf{E}(\mathbf{r}) \cdot \mathbf{j}\,d^3r. \tag{2.9}$$

By integrating over the whole volume

$$P = \int_V \mathbf{j}(\mathbf{r}) \cdot \mathbf{E}(\mathbf{r})\,d^3r \tag{2.10}$$

the total electrical power $P\,[\text{W}]$ is obtained.

2 Fundamentals

In the special case of a thin conductor, following equation can be derived [9, p. 168]

$$P = I \int_C \mathbf{E} \cdot \mathrm{d}r = I \cdot U = R \cdot I^2 = \frac{1}{R} U^2. \tag{2.11}$$

The term

$$P = R \cdot I^2 \tag{2.12}$$

is called power loss. It should be noted that this parameter is proportional to the resistance of a conductor and proportional to the the square of the current.

2.2 Fundamentals of Magnetostatics

In the following sections the mathematically formulation of static magnetic fields is introduced, different magnetostatic groups are listed – respectively the superparamagnetic nanoparticles are assigned – and characteristics of permanent magnets are explained.

2.2.1 Biot–Savart Law

The Ampère's circuital law forms the basis of magnetostatics. It describes the interaction between two electrical conductors (see figure 2.2)

$$F_{12} = \frac{\mu_0 I_1 I_2}{4\pi} \oint_{C_1} \oint_{C_2} \frac{\mathrm{d}\mathbf{r}_1 \times (\mathrm{d}\mathbf{r}_2 \times \mathbf{r}_{12})}{r_{12}^3}. \tag{2.13}$$

Here, the constant μ_0 is the vacuum permeability and amounts to $4\pi \cdot 10^{-7}\,\mathrm{H\,m^{-1}}$. With equation (2.13), the magnetic flux density $\mathbf{B}$ [T] being generated in loop C_2 is defined

$$\mathbf{B}_2(\mathbf{r}_1) = \mu_0 \frac{I_2}{4\pi} \oint_{C_2} \frac{\mathrm{d}\mathbf{r}_2 \times \mathbf{r}_{12}}{r_{12}^3}. \tag{2.14}$$

This equation is called Biot-Savart law and can be extended to arbitrary current densities

$$\mathbf{B}(\mathbf{r}) = \frac{\mu_0}{4\pi} \int \mathrm{d}^3 r' \mathbf{j}(\mathbf{r}') \times \frac{\mathbf{r} - \mathbf{r}'}{|\mathbf{r} - \mathbf{r}'|^3}. \tag{2.15}$$

The Biot-Savart law frames the basis of most magnetic field simulations. Analytical optimizations allow faster algorithms for certain coil forms with elliptical integrals [11].

2.2 Fundamentals of Magnetostatics

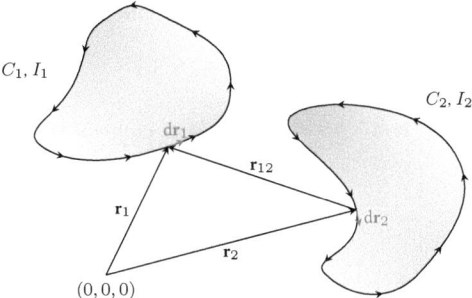

Figure 2.2: Interaction between two current-carrying, closed conductors $C_{1,2}$ with current $I_{1,2}$. The vectors $\mathrm{d}\mathbf{r}_{1,2}$ are infinitesimal parts of the loops with the origin $\mathbf{r}_{1,2}$ [9, p. 168].

2.2.2 Maxwell's Equations

Biot Savart's law given by equation (2.15) can be transformed and it can be shown that the magnetic flux density is a rotation field and source free

$$\operatorname{div}\mathbf{B} = 0 \quad \Leftrightarrow \quad \oint_{S(V)} \mathbf{B}(\mathbf{r}) \cdot \mathrm{d}\mathbf{f} = 0. \tag{2.16}$$

For a detailed derivation see [9, p. 173]. Equation (2.16) is the homogeneous Maxwell equation of magnetostatics. The flux through the surface of a volume $S(V)$ is zero and there are no magnetic monopoles. The inhomogeneous Maxwell equation of magnetostatics is given by

$$\operatorname{rot}\mathbf{B} = \mu_0 \mathbf{j} \quad \Leftrightarrow \quad \int_{\partial F} \mathbf{B} \cdot \mathrm{d}\mathbf{r} = \mu_0 \int_F \mathbf{j} \cdot \mathrm{d}\mathbf{f} = \mu_0 I. \tag{2.17}$$

This equation is called Ampère's circuital law.

So far, only circumstances for the vacuum were described. Thus, an generalization is important.

2.2.3 Magnetic Moment

The magnetic flux density is a rotation field and source free. According to current knowledge there are no magnetic charges [12, p. 291]. However, magnetic charges are introduced due to formal reasons. This simplifies certain physical problems.

First, the magnetization is introduced

$$\mathbf{M} = \frac{\mathrm{d}\mathbf{m}}{\mathrm{d}V}, \tag{2.18}$$

2 Fundamentals

which describes the volume density of the magnetic moment

$$\mathbf{m} = \frac{1}{2} \int d^3 r [\mathbf{r} \times \mathbf{j}(\mathbf{r})]. \tag{2.19}$$

With the magnetization $\mathbf{M}$ [A m^{-1}] and the magnetic moment $\mathbf{m}$ [A m^2] it is possible to calculate the magnetic field of a certain distribution of magnetic dipoles. It should be noted that following relation

$$|\mathbf{j}(\mathbf{r})| = |\mathbf{M}(\mathbf{r})| \tag{2.20}$$

can be inserted in equation (2.15) to calculate the field of e.g. permanent magnets or other magnetic material if the current density respectively the magnetization is equally distributed. The absolute value is also known as remanence.

2.2.4 Magnetostatics in Matter

Since superparamagnetic particles are imaged in MPI and permanent magnets generate the required magnetic fields, they have to be classified concerning their magnetostatic characteristics.

By taking a closer look at Maxwell's equations in matter, equation (2.16) and equation (2.17) can be seized, since averaging does not change these equations in the microscopic case.

However, the current density consists of three different kinds of charges: free charges $\mathbf{j}_f$, polarization charges $\mathbf{j}_p$ and magnetization current density $\mathbf{j}_{mag}$.

The macroscopic, inhomogeneous Maxwell's equation is now defined as

$$\operatorname{rot} \mathbf{B} = \mu_0 \bar{\mathbf{j}}_m = \mu_0 \left(\bar{\mathbf{j}}_f + \bar{\mathbf{j}}_p + \bar{\mathbf{j}}_{mag} \right) = \mu_0 \bar{\mathbf{j}}_f + \mu_0 \dot{\mathbf{P}} + \mu_0 \operatorname{rot} \mathbf{M}. \tag{2.21}$$

The term $\mu_0 \dot{\mathbf{P}}$ is omitted in magnetostatics. Next, the magnetic field

$$\mathbf{H} = \frac{1}{\mu_0} \mathbf{B} - \mathbf{M} \tag{2.22}$$

is introduced. It has the same unit as the magnetization $\mathbf{M}$ [A m^{-1}]. By combining both equations, the inhomogeneous Maxwell's equation can be written as

$$\operatorname{rot} \mathbf{H} = \bar{\mathbf{j}}_f. \tag{2.23}$$

According to this, $\mathbf{H}$ is connected to the free charge and $\mathbf{B}$ to the total current. Now, the magnetic susceptibility χ_m and the relative permeability μ_r are defined linking

$$\mathbf{M} = \chi_m \mathbf{H} \tag{2.24}$$

respectively

$$\mu_r = 1 + \chi_m. \tag{2.25}$$

These physical relations can be used to divide magnetostatic in three different subgroups.

2.2 Fundamentals of Magnetostatics

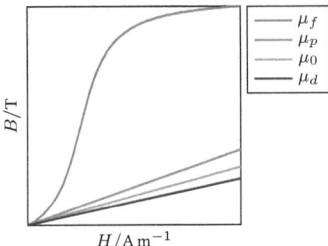

Figure 2.3: Comparism of the permeabilities of ferromagnetic (μ_f), paramagnetic (μ_p), and diamagnetic materials (μ_d) to vacuum (μ_0).

- Diamagnetic materials, whose magnetic permeability μ is slightly smaller than μ_0,
- Paramagnetic materials, whose magnetic permeability μ is slightly greater than μ_0 and
- Ferromagnetic materials, whose magnetic permeability μ is much larger than μ_0.

The behavior of named materials in an external magnetic field is shown in figure 2.3. The plot correlates to

$$\mathbf{B} = \mu_0 H (\chi_m + 1). \tag{2.26}$$

Now, a more detailed summary of the named magnetostatic subgroups is presented.

Diamagnetism

Diamagnetic materials does not feature permanent magnetic dipoles. If this material is brought into a magnetic field, magnetic dipoles are created. This can be describe as an induction effect. The dipole moment is contrariwise aligned to the magnetic field lines. Every material features this characteristic. However, diamagnetism is insignificant in contrast to para- and ferromagnetic effects.

Paramagnetism

Paramagnetic materials rather show a permanent magnetic dipole moment. If an external magnetic field is applied, the dipoles preferably align with the magnetic field lines. Without this field they are randomly aligned. The resulting field – external and dipole field – is slightly larger than the applied one. The effect has a high dependency concerning the temperature. This is described by Curie's law

$$\chi_m(T) = \frac{C}{T}. \tag{2.27}$$

2 Fundamentals

The material specific Curie constant C

$$C = \mu_0 \frac{\mu^2}{n3k_B} \qquad (2.28)$$

depends on the particle density n, the vacuum permeability μ_0, the atomic magnetic moment μ, and the Boltzmann constant k_B with $1.380{,}650{,}4 \cdot 10^{-23}\,\text{J}\text{K}^{-1}$.

If the external magnetic field is switched off, the dipoles randomly re-align as a consequence of thermal fluctuations.

Ferromagnetism

Materials generating a permanent magnetic field without the need of an electric current flow can be allocated in this group. Permanent magnets feature this behavior. The single magnetic moments of a permanent magnet are aligned in one direction (see figure 2.4b).

The domains of a non-magnetized piece of iron are randomly aligned. The magnetic influences of these macroscopic areas cancel out each other (see figure 2.4a). A non-magnetized,

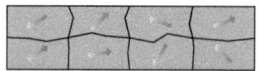
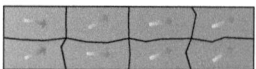

(a) The domains are randomly aligned. The arrows describe the direction of the magnetization. One can assume that a domain behaves like a single magnet.

(b) The domains in a magnet are preferable aligned in one direction.

Figure 2.4: Difference between non magnetized (left) and magnetized (right) materials.

ferromagnetic material can be transformed to a permanent magnet with the help of an external field. Here, the magnetic susceptibility is a complex function of the external field and the temperature

$$\chi_m = \chi_m(T, H). \qquad (2.29)$$

Assumption for this process is the provision of permanent magnetic dipoles that regularly align – without an external field – due to quantum-mechanic interactions below a critical temperature T^*.

In this case, the critical temperature T^* equals T_C. At absolute zero point all moments are parallel aligned. For $0 < T < T_C$ a certain disorder occurs. This disorder increases with temperature. At $T > T_C$ the ferromagnet behaves like a paramagnet. Table 2.1 displays some material specific Curie temperatures. Typical for a ferromagnet is on the one hand a high susceptibility χ_m and on the other hand a great dependency on the materials preparation leading to hysteresis.

2.2 Fundamentals of Magnetostatics

Substance	Fe	Co	Ni	Gd	EuO	CrBr$_3$	Nd$_3$Fe$_{14}$B
T_C/K	1,043	1,393	631	290	69	37	583

Table 2.1: Curie temperatures of some materials [9, p. 188].

Hysteresis Hysteresis characterizes a system whose variable output parameter is not just dependent on the input parameter but on the course, too. Ferromagnetic materials such as iron, cobalt and nickel feature this named characteristic.

If an external magnetic field is applied to an unmagnetized piece of iron (see figure 2.5), it is magnetized corresponding the

- initial magnetization curve (a)

and goes into saturation (b). If the field is switched off, the magnetization of the material does not completely disappear. This magnetization is called

- remanence (c)

and can be cancelled out with an opposing magnetic field being referred as

- coercive force (d).

The remanence (c) characterizes a permanent magnet. Strong magnets feature a high remanence. One of the strongest, available materials for permanent magnets is neodymium iron boron. A remanence of up to $1.4\,\text{T}$ can be achieved.

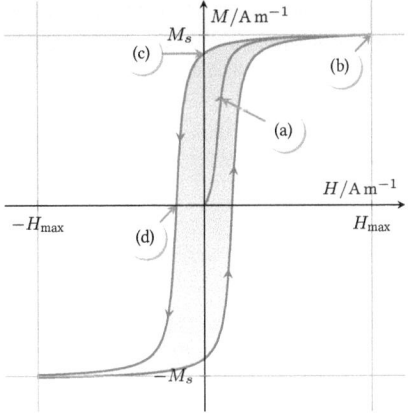

Figure 2.5: Hysteresis of ferromagnetic material (from [13, p. 12]).

2 Fundamentals

Superparamagnetism

Iron oxide nanoparticles are used as tracer in MPI and feature multiple magnetostatic attributes. They belong to the ferromagnetic materials but feature – if it comes to low concentrations – paramagnetic characteristics with high magnetic susceptibility χ_m [14]. Basic requirement is that the particle core is so small that it behaves like a single-domain particle. Then, one particle operates as a single magnetic moment. The core consists of iron oxide such as magnetite (Fe_3O_4). This is called superparamagnetism [15].

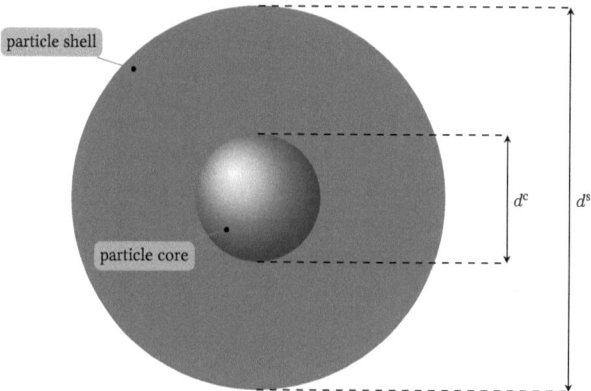

Figure 2.6: Schematic drawing of a super-paramagnetic nanoparticle. The particle core is surrounded by a shell to avoid agglomeration [16, p. 30].

Besides the core, the particles consist of a neutral shell avoiding agglomeration. This shell can for instance consist of dextran or carboxydextran and enables biocompatibility [17]. Further shell optimizations may extend the application area in the clinical sector. A commonly used tracer is Resovist from Bayer-Schering that is also used in magnetic resonance imaging (MRI) [18].

The size of the shell avoids magnetic influences between the particles. Typical dimensions for superparamagnetic particles are core diameters (d^c) between 1 nm - 30 nm and hydrodynamic diameters (d^s) between 40 nm - 400 nm. Figure 2.6 images the core respectively the shell of a particle.

2.3 Electrodynamics

Besides static magnetic fields, MPI uses also alternating magnetic fields. To complete Maxwell's equations, the dynamic case has also to be covered. Here, electrostatic and magnetostatic phenomena cannot be discussed separately.

2.3.1 Faraday's Law

A current is generated in a conductor by moving it through a magnetic field. This is called electromagnetic induction and was discovered by Faraday in 1861 [19]. It can be shown that the electromotive force is proportional to the magnetic flux

$$\oint_C \mathbf{E} \cdot d\mathbf{r} = -\frac{d}{dt} \int_{F_C} \mathbf{B} \cdot d\mathbf{f} \qquad (2.30)$$

and analogous

$$\operatorname{rot} \mathbf{E} = -\dot{\mathbf{B}} \qquad (2.31)$$

2.3.2 Maxwell's Extension

Previously, the magnetic field $\mathbf{H}$ was introduced by omitting the term $\dot{\mathbf{P}}$ in section 2.2.4, since only magnetostatic phenomena were considered. Maxwell generalized equation (2.23) by introducing the displacement current $\dot{\mathbf{D}}$. Consequently, the following relation applies

$$\operatorname{rot} \mathbf{H} = \mathbf{j} + \dot{\mathbf{D}}. \qquad (2.32)$$

2.4 Maxwell's Equations

The four fundamental Maxwell's equations were introduced in section 2.2.4 and section 2.3. Thus, in the following the complete Maxwell's equations are summarized by adding two extensions for the sake of completeness [19, 20, 21]:
Gauss's law

$$\operatorname{div} \mathbf{D} = \varrho \qquad (2.33)$$

Gauss's law for magnetism

$$\operatorname{div} \mathbf{B} = 0 \qquad (2.34)$$

Faraday's law of induction

$$\operatorname{rot} \mathbf{E} + \dot{\mathbf{B}} = 0 \qquad (2.35)$$

Ampre's circuital law

$$\operatorname{rot} \mathbf{H} - \dot{\mathbf{D}} = \mathbf{j} \qquad (2.36)$$

Two vectorial and two scalar equations are available for five vectorial values (see table 2.2).

2 Fundamentals

Name	Symbol	SI Unit
Electrical field	E	V m^{-1}
Electric displacement field	D	A s m^{-2}
Magnetic field strength	H	A m^{-1}
Magnetic flux density	B	T
Total current density	j	A m^{-2}
Total charge density	ϱ	C m^{-3}

Table 2.2: Vectorial and scalar values that appear in Maxwell's equations.

Additional equations link the magnetic flux density **B** to the magnetic field strength **H** (as already discussed in equation (2.22)), respectively the electric displacement field **D** to the electric field **E**. Hence, the polarization **P** [C m^{-2}] is introduced

$$\mathbf{B} = \mu_0(\mathbf{H} + \mathbf{M}), \tag{2.37}$$
$$\mathbf{D} = \epsilon_0 \mathbf{E} + \mathbf{P}. \tag{2.38}$$

The constant ϵ_0 is the vacuum permittivity and amounts to $8.854 \cdot 10^{-12}$ F m^{-1}. The magnetization **M** as well as the polarization **P** are usually time and location dependent.

2.5 Hall Probe

An Hall probe facilitates measuring magnetic fields. Since it is used to evaluate the coil configuration from section 3.1, a brief introduction is presented to discuss its functionality [22, p. 935].

2.5.1 Hall Effect

By inserting a conductor in an homogeneous magnetic field, the charges exhibit a force caused by the magnetic field. Figure 2.7 demonstrates this. The electrons are flowing through the conductor from left to right. The magnetic field is perpendicular to the drawing plane and applies a force

$$\mathbf{F_B} = -e\mathbf{v}_d \times \mathbf{B} \tag{2.39}$$

to the electrons. The elementary charge e is $1.602{,}176{,}565 \cdot 10^{-19}$ C and $\mathbf{v}_d$ the drift velocity of the electrons. This causes a potential difference of the electrons and correspondingly an electric field $\mathbf{E}_H$. This process progresses as long as the electrical field $\mathbf{E}_H$ applies a force $e\mathbf{B}_H$ to the charges that has the same absolute value as the force of the magnetic field but with opposite sign. This balance is called Hall effect. The created potential difference U_H is called Hall voltage.

2.5 Hall Probe

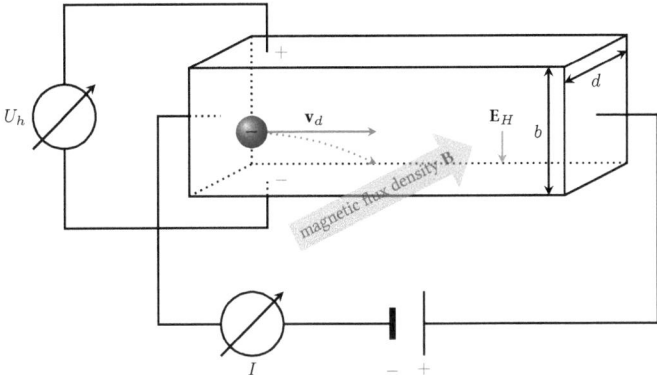

Figure 2.7: Visualization of the Hall effect. By considering the physical current direction, negative charges move in the direction of the current (here: to the right).

The electrical field E_H (Hall field), being generated by separation of charge, is directed downwards. In the next steps the absolute value of the electrical field and the magnetic flux density is taken into account. At equilibrium applies

$$eE_H = ev_d B. \tag{2.40}$$

This means that the electrical force is equal to the magnetic force. By assuming a long and thin electrical conductor and an homogeneous electrical field, following relation can be verified

$$U_H = E_H b \tag{2.41}$$
$$= v_d B b, \tag{2.42}$$

whereat b is the thickness of the conductor. With $v_d = \frac{I}{nFq}$ (see equation (2.2)) the drift velocity can be replaced:

$$U_H = \frac{I}{nFq} Bb \tag{2.43}$$
$$= \frac{I}{nbdq} Bb \tag{2.44}$$
$$= \frac{IB}{ndq}. \tag{2.45}$$

Hence, the voltage U_H can be simplified

$$U_H = A_H \frac{IB}{d}. \tag{2.46}$$

The Hall constant A_H is the inverse of the particle density n times the charge q.

2 Fundamentals

Since the amplitude of the Hall voltage is proportional to the magnetic flux density, the magnetic flux density can easily be calculated

$$B = \frac{U_H d}{A_H I}. \tag{2.47}$$

The electrical conductor is called Hall probe and can be calibrated with an known field strength.

2.6 Magnetic Particle Imaging

Magnetic Particle Imaging (MPI) is a new imaging modality capable of imaging the distribution of superparamagnetic nanoparticles with high spatial and temporal resolution. It was first published by Bernhard Gleich and Jürgen Weizenecker in 2005 [1]. Since then, steady progress in imaging techniques, image quality, hardware components, and simulation studies could be noted. In 2009, first in-vivo imaging could be realized [2]. A beating mouse heart was imaged with a resolution of 40 frames per second. The high spatial resolution allowed differentiating the heart ventricles. This method promises high potential since the mouse heart beats with a frequency of about 4 Hz or 240 beats per minute. The examined field of view (FOV) was 20.4 mm × 12.0 mm × 16.8 mm.

In the following a short overview and comparison of further imaging systems shall be given, in order to name some future clinical applications.

Figure 2.8 gives an overview of imaging systems already being used in clinical practice and the comparison to the potential of prospective MPI. Important parameters for a clinical device are spatial resolution, measurement speed and sensitivity. Furthermore, the radiation exposure is a crucial criterion. Ionization is not just harmful for the patient but all the more for the radiologist being daily exposed to radiation.

Computed tomography (CT) is a fast method with high spatial resolution and is commonly used in trauma diagnostic [23, 24, 25]. The disadvantage is a certain radiation exposure. Magnetic resonance imaging (MRI) offers great soft tissue contrast and good spatial resolution but requires a long measurement time [26, 27]. Real-time imaging is only possible with low resolution and respiratory or rather cardiac triggering [28]. Positron emission tomography (PET) [29, 30] and single-photon emission computed tomography (SPECT) [31] are quantitative methods with a high sensitivity. They are particularly used in the field of functional imaging. MPI does combine high spatial and temporal resolution and a good sensitivity without having ionizing characteristics.

Note that this is only a schematic chart. MPI does not – in contrast to MRI or CT – image anatomical structures but only the tracer. It is a quantitative method. Hence, future considerations discuss a fusion between e.g. MPI and MRI.

Future clinical applications could be the sentinel lymph node biopsy if the suspicion of breast cancer exists [32, 33]. Further applications are especially scenarios where a fast and dynamic

2.6 Magnetic Particle Imaging

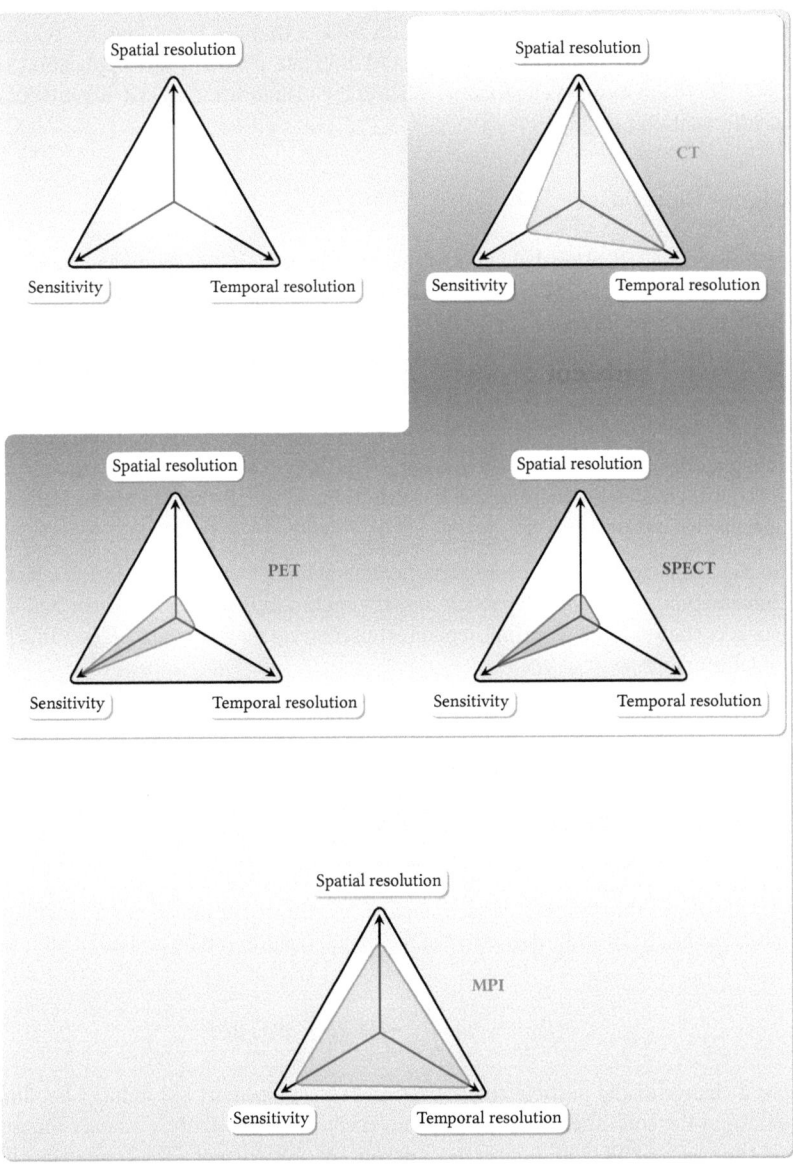

Figure 2.8: Comparison between different imaging systems being used in clinical practice [16].

2 Fundamentals

imagining process is necessary. Here, diagnostics of coronary vessels and possible vasoconstriction and angiography could be possible [3]. Since X-ray angiography, CT angiography and MRI angiography use iodine or gadolinium tracers that are hazardous for patients with Chronic Kidney Disease, MPI angiography could motivate a patient safe application scenario [34, 35, 36]. There are also developments realizing the visualization of instruments for cardiovascular intervention [37, 38].

2.6.1 Signal Generation

The non-linear magnetization behavior of superparamagnetic nanoparticles frames the basis for signal generation in MPI. The magnetization of the particles can be described by the Langevin function (see figure 2.9 (b)) as a function of the magnetic field

$$M(H) = \begin{cases} cm(\coth(\xi) - \frac{1}{\xi}) & H \neq 0 \\ 0 & H = 0 \end{cases} \quad \text{with} \quad \xi = \frac{\mu_0 m H}{k_B T}. \tag{2.48}$$

Here, c characterizes the particle concentration, T the temperature, k_B the Boltzmann constant, $m = \frac{1}{6}\pi D^3 M_{\text{sat}}$ the magnetic moment in saturation, D the particle diameter, and M_{sat} the saturation magnetization.

Now, the idea is to use the non-linear magnetization behavior and the saturation effect to gain spatial information. But first, the particle signal generation is discussed. Therefore, a certain particle concentration is put in a time dependent oscillating magnetic field $H(t)$. This field has a sinusoidal course

$$H(t) = H_0 \sin(2\pi f t). \tag{2.49}$$

Hence, the Langevin function (see equation (2.48)) is time dependent and can be written as

$$M(t) = \begin{cases} cm(\coth(\xi) - \frac{1}{\xi}) & H \neq 0 \\ 0 & H = 0 \end{cases} \quad \text{with} \quad \xi = \frac{\mu_0 m H_0 \sin(2\pi f t)}{k_B T}. \tag{2.50}$$

Figure 2.9 illustrates this process. Now, the particles magnetization shows a modulated progression. Accordingly, the time derivative of the magnetization can be measured with a receive coil

$$u(t) = -\mu_0 \int_S \frac{\mathrm{d}}{\mathrm{d}t} \mathbf{M}(\mathbf{r}, t) \cdot \mathbf{p}(\mathbf{r}) \, \mathrm{d}A. \tag{2.51}$$

A temporal change of the particle magnetization $\mathbf{M}(\mathbf{r}, t)$ (see part (c)) induces a voltage $u(t)$ (see part (d)) in the coil. The vector $\mathbf{p}(\mathbf{r})$ is the receive coil sensitivity – exactly the magnetic field that the receive coil would generate if driven by unit current. The frequency spectrum of the signal shows the odd harmonics in part e). Higher harmonics in the yellow frame are characteristic for the particles. Harmonic f_1 contains the drive field frequency and also information from the particles. It can be noted that in a realistic situation, noise influences would superimpose with the particle signal.

2.6 Magnetic Particle Imaging

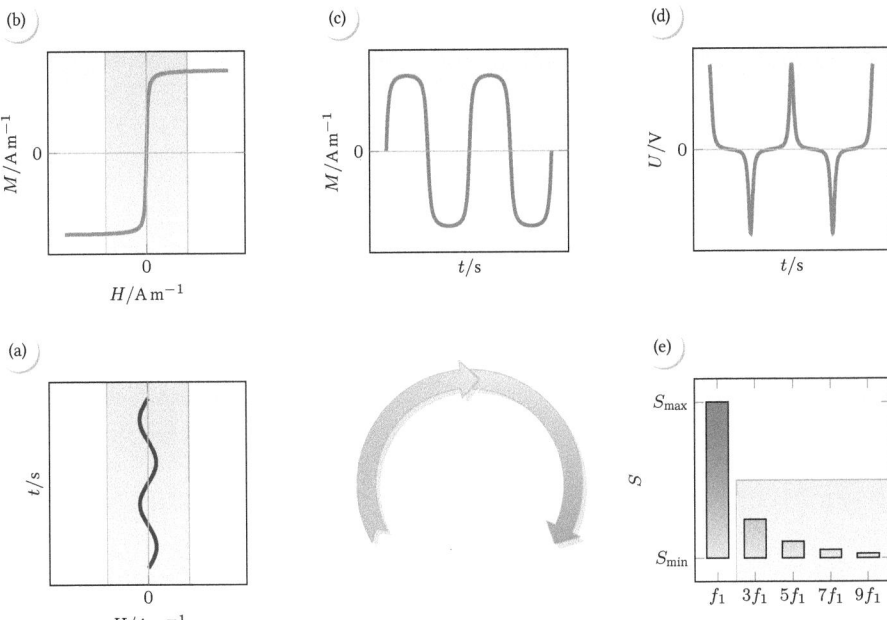

Figure 2.9: This visualization describes the process of signal generation in MPI: a time dependent sinusoidal magnetic field (a) excites the nanoparticle in the yellow area (b). The particles magnetization changes over time (c) and induces a voltage in the receive coil (d). The induced signal has a characteristic frequency spectrum (e). The frequency f_1 is the addition of particle signal and sinusoidal excitation. However, part (d) only shows the particle signal for better visualization.

Now, it is theoretically possible to generate a particle signal, respectively detect particles. But, the signal does not include any spatial information. Hence, in the following it is explained how particle signals can be suppressed (see figure 2.10). An homogeneous offset field H_{offset} is added to the oscillating field $H(t)$ (see part (a)). Now, the particles are excited in the saturation area (part (b)) and hence, the magnetization does not change over time (part (c)). According to this, no signal is induced in the receive coil (part (d)) and the frequency spectrum is empty and does not contain any higher harmonics.

The idea is to suppress the particles signal except in the region of interest. This can be realized by a gradient field that is e.g. generated by two permanent magnets with opposite magnetization direction (see figure 2.11) or a Maxwell coil pair. This field is also called selection field.

A field characteristic is the field free point (FFP) at the center between both magnets – a specific area where both magnetic fields of the permanent magnets cancel out each other. From

2 Fundamentals

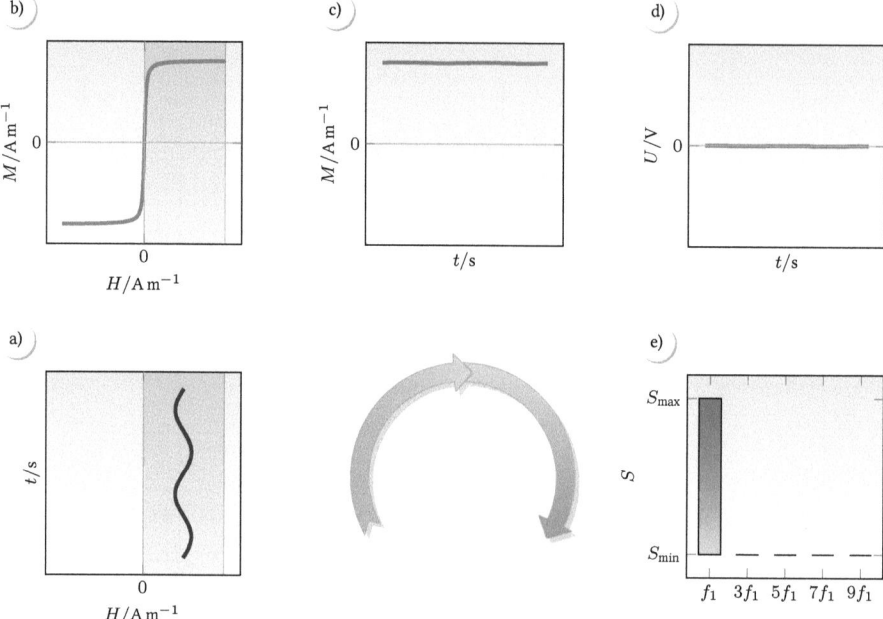

Figure 2.10: This visualization describes the process of signal suppression in MPI: a time dependent sinusoidal magnetic field that is superimposed with a magnetic offset field (a) excites the nanoparticle in the yellow area (b). Since the particle is already saturated, the magnetization does not change significantly over time (c). Thus, almost no voltage is induced in the receive coil (d) and accordingly the frequency spectrum does only contain the frequency of the excitation field (e). The frequency f_1 is the addition of particle signal and sinusoidal excitation. However, part (d) only shows the particle signal for better visualization.

this point the magnetic field rises in all directions. By moving the FFP either mechanically or electronically a certain field of view (FOV) can be scanned. Whenever the FFP crosses a particle concentration, a signal is generated and measured in the receive coils.

Signal Generation Using a Field Free Line

The FFP has just been introduced as a way to spatially encode a particle signal. An advanced encoding scheme uses a field free line (FFL) [4]. It is possible to extend the FFP to an FFL by dispersing it. This is visualized in figure 2.12. Analogous to the FFP, the FFL is moved through the FOV. Usually this is done by moving the FFL over the FOV. Whereas the direction is perpendicular to the FFLs course. Simultaneously, the FFL is rotated about its center point.

2.6 Magnetic Particle Imaging

Figure 2.11: A gradient field generated by two permanent magnets with opposing magnetization.

FFL shift and rotation form a complex and continuous trajectory. An alternative method is to use a static FFL and rotate the object to be examined [39].

An aim in MPI is to increase resolution by increasing the gradient (see section 2.6.2). But, higher gradients minimize the number of particles contributing to the signal which means a loss of sensitivity. This simultaneous encoding scheme can potentially increase the sensitivity by one order of magnitude – in comparison to conventional FFP imaging [7].

For a detailed description of FFL generation, coil configuration and power loss optimization see section 3.1.

In the next section the relation between gradient field, particle characteristic and resolution is described. Afterwards, the reconstruction process of the signal is analyzed.

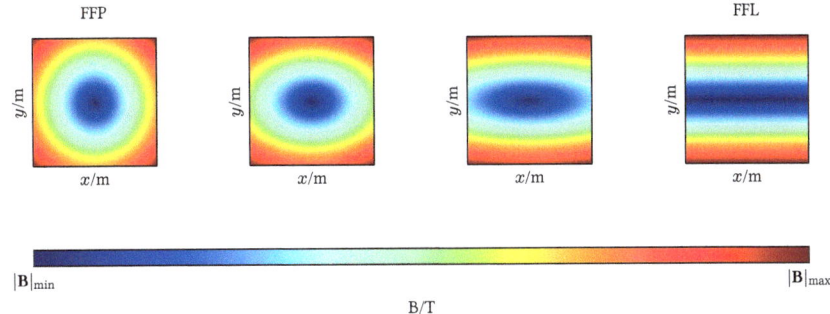

Figure 2.12: Demonstration how an FFP is stretched to an FFL. It is clear that due to the larger area which the FFL covers, more particle are excited. This leads to a higher sensitivity.

2 Fundamentals

2.6.2 Resolution

An important property of MPI is the high spatial resolution which is directly connected to the gradient field.

Figure 2.13 shows two gradient fields: the green one with a high gradient and the blue one with a low gradient. By comparing both fields and searching for the intersection line that separates

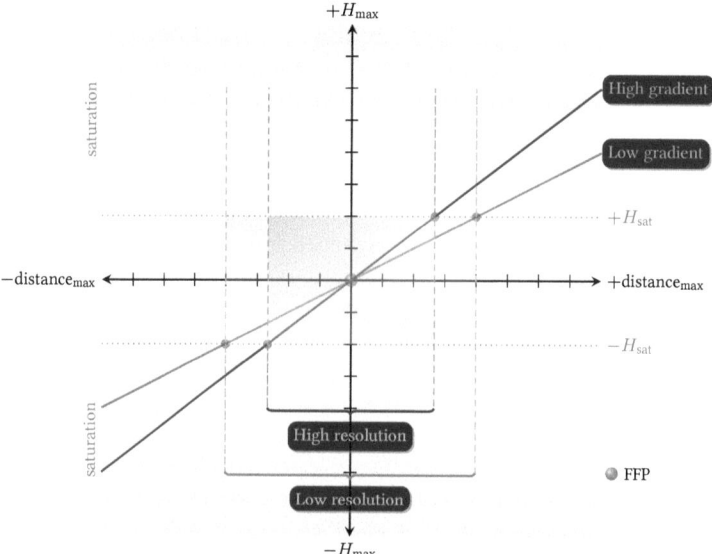

Figure 2.13: Schematic drawing of two different gradient fields and their effect on the resolution. The x-axis marks the location respectively the y-axis the magnetic field strength. A higher gradient minimizes the FFP area and consequently increases the resolution.

saturated and non saturated area, it is obvious that higher gradients produce a smaller non saturated area and a higher resolution. This characteristic is crucial for MPI and can also be derived mathematically.

The derivative of the Langevin function (see figure 2.14, (b)) from equation (2.48) can be written as

$$\frac{d}{d\xi}M(H) = cm\left(\frac{1}{\xi^2} - \frac{1}{\sinh^2(\xi)}\right). \tag{2.52}$$

This function is plotted in figure 2.14 b) and c). According to [14], the full width at half maximum (FWHM) approximates the achievable resolution: it can be calculated numerically as $\Delta\xi = 4.16$. If the magnetic moment m of the particles and the gradient of the selection field are

2.6 Magnetic Particle Imaging

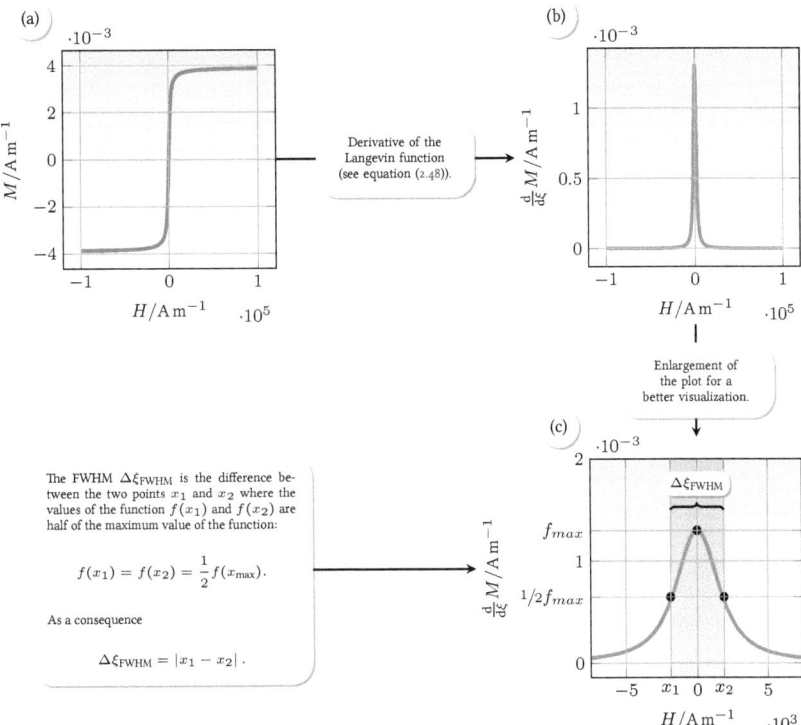

Figure 2.14: Visualization of the Langevin function (a) and its derivative (b). Part (c) is a clipping from (b) and marks the FWHM.

known, the spatial resolution can be approximated with this equation

$$\Delta x \approx \frac{k_B T}{\mu_0 m G} \Delta \xi_{\text{FWHM}}. \tag{2.53}$$

Hence, the resolution of MPI is inverse proportional to the gradient strength of the selection field and since the magnetic moment $m = \frac{\pi}{6} M_{\text{sat}} D^3$ is exponentially linked to the particle diameter D, a doubling of the particle diameter would result in an eight times higher resolution.

These facts are only idealized mathematical assumptions. In practice one has to deal with noise, regularization during the reconstruction process, and non-ideal particles that lower the resolution.

However, increasing the gradient can have limitations with respect to the human body. Certain gradient and frequency combinations cause peripheral nerve stimulation (PNS) and tissue heating, known as specific absorption rate (SAR) [40, 41, 42].

2.6.3 Reconstruction

Two kinds of spatial encoding schemes have been mentioned so far that are used in MPI: the FFP and the FFL encoding. Thus, corresponding reconstruction techniques have to be implemented. A common reconstruction technique uses a system matrix to reconstruct the signal to an image [1]. This system matrix can be used for FFP as well as FFL trajectories.

Furthermore, a Radon-based reconstruction method uses the field homogeneity of the FFL to facilitates fast imaging [7].

It can be noted that in addition to these techniques, a further reconstruction scheme exists that does not need a System Matrix and is called x-Space reconstruction. For more information see [39, 43, 44, 45, 46].

Reconstruction Using a System Matrix

Analogous to other imaging systems – e.g. CT [23, p. 201 ff.] – the reconstruction process can be modeled as a linear system of equations. In MPI a matrix vector formulation is existent

$$\mathbf{Sc} = \hat{\mathbf{u}} \tag{2.54}$$

with the system matrix $\mathbf{S} \in \mathbb{C}^{M \times N}$, the unknown particle distribution $\mathbf{c} \in \mathbb{C}^N$ and the measured signal $\hat{\mathbf{u}} \in \mathbb{C}^M$. In the ideal case, the measured signal is noiseless, but since additional signal generation is not avoidable, the signal $\tilde{\mathbf{u}} = \hat{\mathbf{u}} + \eta$ is measured. Here, η is noise. Thus, the system

$$\mathbf{Sc} \approx \tilde{\mathbf{u}} \tag{2.55}$$

has to be solved.

In contrast to the measured data, the system matrix $\mathbf{S}$ is unknown in MPI. The next section describes the determination of the system matrix. If the system matrix is known there are several possibilities to reconstruct the image.

Since noise influences the measured data, the uniqueness of a solution cannot be guaranteed. That is why often a weighted least squares approach is used being robust as against noise [16, p. 88]. An algorithm determining weighted least squares solution is e.g. a singular value decomposition (SVD). However, this method is very time and memory consuming. Faster algorithms such as the conjugate gradient and Kaczmarz method use an iterative approach. The interested reader is referred to [47, 1, 48, 49].

2.6 Magnetic Particle Imaging

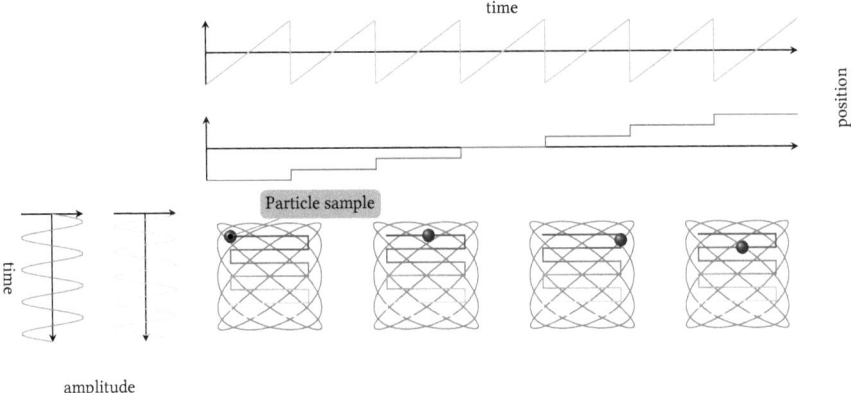

Figure 2.15: The system matrix is gained by doing a complete measurement of every discrete particle sample position. Usually the FOV is partitioned into a grid where a grid position corresponds to one particle sample position and accordingly a full measurement.

Measurement-based System Function The system matrix $\mathbf{S}$ – being necessary to reconstruct the image – can be obtained as follows. First, a particle sample with known concentration c_0, geometry and volume ΔV has to be realized. Usually it is mounted on a robot. Subsequently, the sample is moved to a position $\mathbf{r}_n$ on the FOV. Figure 2.15 demonstrates the sample (green dot) and the positions $\mathbf{r}_n$ (green faded path). Now, the induced signal $u_l(t)$ for a complete scan is measured. The red trajectory visualizes the scanning path. The frequency components $\hat{\mathbf{u}}_{l,k}$ can be calculated afterwards.

This process has to be repeated for all positions $\mathbf{r}_n$, $n = 0, \ldots, N-1$.

In general an appropriate volume size of the sample has to be considered. On the one hand a small sample features a high resolution of the grid, on the other hand signals SNR and accordingly system matrices SNR decreases. So there has to be a trade-off between SNR and resolution.

A critical problem is the long lasting measurement time that can utilize several hours – or even longer for larger FOVs – and high memory requirements depending on the FOVs size and grid resolution.

Further methods to obtain the system matrix using e.g. a model-based approach or a particular combination of both are described in [50, 51].

2 Fundamentals

Radon-based Reconstruction

The Radon-based reconstruction can only operate on data generated with an FFL. The reconstruction process is very similar to the backprojection technique described in [23, p. 175 ff.]. Johann Radon introduced the Radon transform in 1917 [52]. The following mathematical derivation is summarized from [7].

First, a static coordinate system and a rotational coordinate system belonging to the FFL (see figure 2.16) are defined. The FFL is rotatable with an angle γ and shiftable in ξ-direction. The parameter $\mathbf{d}_{\text{FFL}}^{\gamma} = (\cos(\gamma), \sin(\gamma), 0)^T$ describes the minimum distance between FFL and point of origin. The FFL field $\mathbf{H}_S^{\gamma}$ can be described by

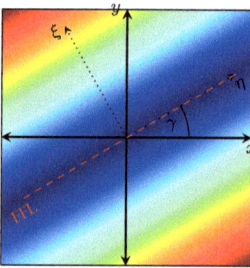

Figure 2.16: Static, cartesian coordinate system and rotational coordinate system of the FFL.

$$\mathbf{H}_S^{\gamma}(x, y) = (Gx \sin(\gamma) - Gy \cos(\gamma)) \begin{pmatrix} -\sin(\gamma) \\ \cos(\gamma) \\ 0 \end{pmatrix}, \qquad (2.56)$$

where G is the gradient. This completely defines the selection field. Note that the gradient field features a constant gradient. This constant gradient is a condition for the following reconstruction technique. Besides, the linearity of the FFL has to be guaranteed. An FFL shift to the FOV edges may not lead to distortion. Regarding practical implementation, this field quality has to be realized (see section 3.1).

Besides the rotation, the FFL is moved back and forth to excite the particles and scan the FOV. A drive field inherits this task

$$\mathbf{H}_D^{\gamma}(t) = A\Delta(t) \begin{pmatrix} -\sin(\gamma) \\ \cos(\gamma) \\ 0 \end{pmatrix}. \qquad (2.57)$$

The factor A is the maximum field amplitude [T μ_o^{-1}] and $\Delta(t)$ is usually a sinusoidal function

2.6 Magnetic Particle Imaging

$\Delta(t) = \cos(2\pi f_0 t)$ which is time dependent. By summing up both fields

$$\mathbf{H}^\gamma(x, y, t) = \mathbf{H}^\gamma_S(x, y) + \mathbf{H}^\gamma_D(t) \tag{2.58}$$

$$= (A\Delta(t) + Gx\sin(\gamma) - Gy\cos(\gamma)) \begin{pmatrix} -\sin(\gamma) \\ \cos(\gamma) \\ 0 \end{pmatrix}, \tag{2.59}$$

the total magnetic field is obtained. An area within a range of $\left[-\frac{A}{G}, \frac{A}{G}\right]$ can be covered. Now, the mean magnetic moment of the particles can be written as a function of the field $\mathbf{H}^\gamma$

$$\mathbf{m}(x, y, t) = m\left(A\Delta(t) + Gx\sin(\gamma) - Gy\cos(\gamma)\right) \begin{pmatrix} -\sin(\gamma) \\ \cos(\gamma) \\ 0 \end{pmatrix}. \tag{2.60}$$

As mentioned in equation (2.51), a change of the particle magnetization induces a voltage in the receive coil. The total magnetization is the product of the particle concentration $c(x, y)$ times the magnetic moment $\mathbf{m}(x, y, t)$

$$\mathbf{M}(x, y, t) = c(x, y) \cdot \mathbf{m}(x, y, t). \tag{2.61}$$

Now, the induced voltage can be written as following

$$u^\gamma(t) = -\mu_0 \int_{\mathbb{R}^2} c(x, y) \frac{\partial}{\partial t} \mathbf{m}(x, y, t) \cdot \mathbf{p}(x, y) \, dx \, dy. \tag{2.62}$$

Furthermore, the induced signal is related to the radon transform of the particle concentration

$$u^\gamma(t) = q^\gamma A \Lambda'(t) (\tilde{m} * \mathcal{R}(c)(\gamma, \cdot)) \left(\frac{A}{G} \Lambda(t)\right) \tag{2.63}$$

with

$$\mathcal{R}(c)(\gamma, \xi) = \int_{\mathbb{R}} c\left(\eta \cos(\gamma) - \xi \sin(\gamma), \eta \sin(\gamma) + \xi \cos(\gamma)\right) d\eta. \tag{2.64}$$

The convolution kernel

$$\tilde{m} = -\mu_0 m(Gx) \tag{2.65}$$

is the derivative of the magnetic moments absolute value and

$$q^\gamma = \begin{pmatrix} -\sin(\gamma) \\ \cos(\gamma) \\ 0 \end{pmatrix} \cdot \mathbf{p} \tag{2.66}$$

is the dot product between the direction of the FFL movement and the receive coil sensitivity. The derivation can be found in [7].

2 Fundamentals

Since the FFL scans the FOV with a radial trajectory, the signal has to be normalized with respect to FFL velocity. This can be done by dividing the signal $u^\gamma(t)$ through the derivative of the excitation function

$$s^\gamma(t) = \frac{u^\gamma(t)}{A\Lambda'(t)} \tag{2.67}$$

$$= (\tilde{m} * \mathcal{R}(c)(\gamma, \cdot))\left(\frac{A}{G}\Lambda(t)\right). \tag{2.68}$$

At points where the FFL changes its direction, the velocity is zero. Hence, one has to dismiss the signal at these named points.

The next step is to transform the signal $s^\gamma(t)$ to the spatial interval $[-\frac{A}{G}, \frac{A}{G}]$. The coordinate transform

$$\xi_{\text{FFL}}(t) = \frac{A}{G}\Lambda(t) \Leftrightarrow T(\xi_{\text{FFL}})\Lambda^{-1}\left(\frac{G}{A}\xi_{\text{FFL}}\right) \tag{2.69}$$

is helpful to make the signal $s^\gamma(t) \Leftrightarrow \tilde{s}(\xi_{\text{FFL}})$ spatial dependent

$$\tilde{s}^\gamma(\xi_{\text{FFL}}) = s^\gamma\left(\Lambda^{-1}\left(\frac{G}{A}\xi_{\text{FFL}}\right)\right). \tag{2.70}$$

The Fourier slice theorem states that a convolution in spatial space is a multiplication in Fourier space. According to this, the following relation can be used

$$\hat{s}^\gamma(v) = \hat{\mathcal{R}}(c)(\gamma, v)\hat{m}(v) \tag{2.71}$$

with

$$\hat{\mathcal{R}}(c)(\gamma, v) = \mathcal{F}(\mathcal{R}(c)(\gamma, \xi)) \tag{2.72}$$

to reconstruct the Radon data.

In a practical application noise has to be considered that is added to the signal. Therefore, a Wiener filter [53, 54] is often used to damp frequency components with low signal-to-noise ratio (SNR)

$$\tilde{\mathcal{R}}(c)(\gamma, \xi) = \mathcal{F}^{-1}\left(\frac{\hat{s}^\gamma(v)}{\hat{m}(v)}\left(\frac{|\hat{m}(v)|^2}{|\hat{m}(v)|^2 + \frac{1}{\text{SNR}(v)}}\right)\right). \tag{2.73}$$

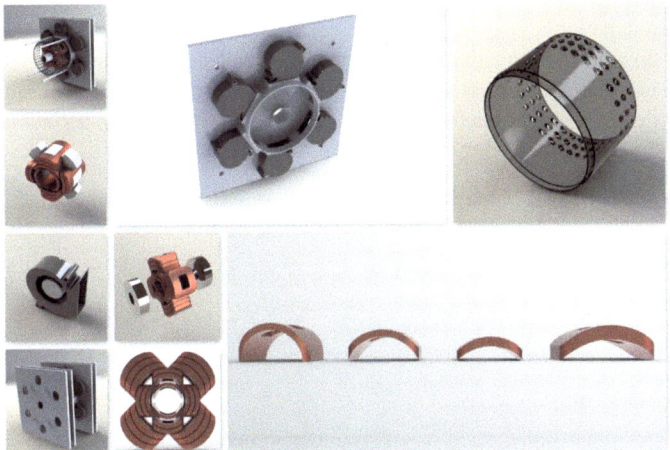

Chapter 3

Materials and Methods

In the following chapter a detailed description of coil design, manufacturing process, and assembling is depicted. Furthermore, the scanner case design, uniting coil configuration, permanent magnets, and cooling system is presented. Finally, it is dwelled on the measurement procedure of the FFL field and its required preparations.

3 Materials and Methods

3.1 Introduction and Coil Configuration

The concept of an FFL was first introduced in [4]. A simulation study showed the feasibility and much higher image quality than comparable scans with an FFP. 16 Maxwell coil pairs, aligned on a circle with a diameter of 1 m were needed to generate the selection field. This entails an immense power loss which would be a thousand times higher than an FFP scanner with same gradient strength and equal size [5]. An appropriate cooling system would be very complex and not feasible.

Further simulations could reduce the amount of selection field coils [55]. Only three respec-

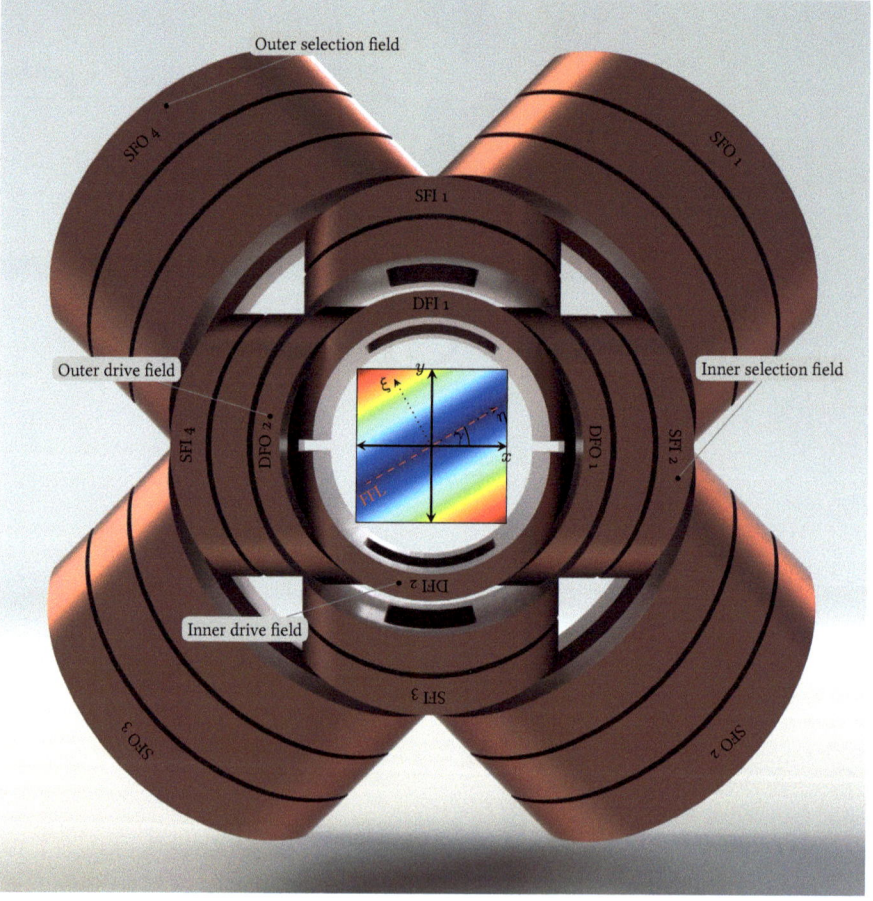

Figure 3.1: The picture shows the FFL generator consisting of 24 coils. These coils build the inner and outer selection field array and inner and outer drive field array.

3.1 Introduction and Coil Configuration

tively four Maxwell coil pairs are needed. Whereat three coils feature less power loss and four coils improve magnetic field quality. One has to decide wether it is the power loss or the field quality that should be optimized. The power loss was in the range of 3.3 to 6.9 times as much as a comparable FFP scanner. Since field quality – the homogeneity of the field along the FFL – is of high importance for an imaging device, four selection field coil pairs proved to be optimal.

The first experimentally generated rotated and translated FFL field was presented in [6]. This was realized with four selection field coil pairs and a gradient strength of $0.25\,\text{T}\,\text{m}^{-1}\,\mu_0^{-1}$ was achieved. The power loss could be minimized to $42.32\,\text{W}$ for pure rotation. An additional translation added up to $49.06\,\text{W}$.

The latest simulation study suggests curved rectangular coils to generate the FFL [8]. This work shows a five times higher field quality and an almost four times lower power consumption of the curved rectangular coils in comparison to circular coils. The coil concept shown in figure 3.1 is based on this simulation and is evaluated in this thesis.

In the following, the coil configuration is introduced and described. The basic coil concept is illustrated in figure 3.1. It consists of 24 coils that can be assigned to four different coil types (see figure 3.2). The permanent magnets being aligned on the z-axis (perpendicular to the paper plane) are not shown here due to a better visualization (see figure 3.9 for a side view). The symmetrical design includes two selection field rings allowing for generation and rotation of the FFL field and two perpendicular drive field arrays shifting the FFL field.

The selection field rings are split up into an outer and inner selection field ring. Both contain four coil arrays: three coils form one outer selection field array, two coils form one inner selection field array. The idea behind this array concept is to facilitates extra slits for air cooling. These slits increase the coil surface and consequently the cooling efficiency.

One drive field array has the functionality of a Helmholtz coil and generates an homogeneous field. Inner and outer drive field coils are perpendicular aligned. The inner coils shift the FFL in y-direction and the outer coils in x-direction. A combination enables a movement in any direction in the x/y-plane. Figure 3.2 and table 3.1 picture the four coils and their geometry.

	DFInner in mm	DFOuter in mm	SFInner in mm	SFOuter in mm
SideA	60.00	60.00	60.00	60.00
SideB	46.90	46.90	43.84	63.64
InnerSideA	27.50	27.50	31.00	21.00
InnerSideB	24.00	24.60	20.70	31.04
InnerCurvature	20.00	25.50	31.00	45.00
OuterCurvature	23.50	29.00	31.00	45.00
InnerFillet	3.00	3.00	3.00	3.00
OuterFillet	10.00	10.00	12.00	12.00
Thickness	4.40	4.50	6.00	9.00

Table 3.1: Geometric values of selection and drive field coils. Figure 3.2 illustrates the used shortcuts.

3 Materials and Methods

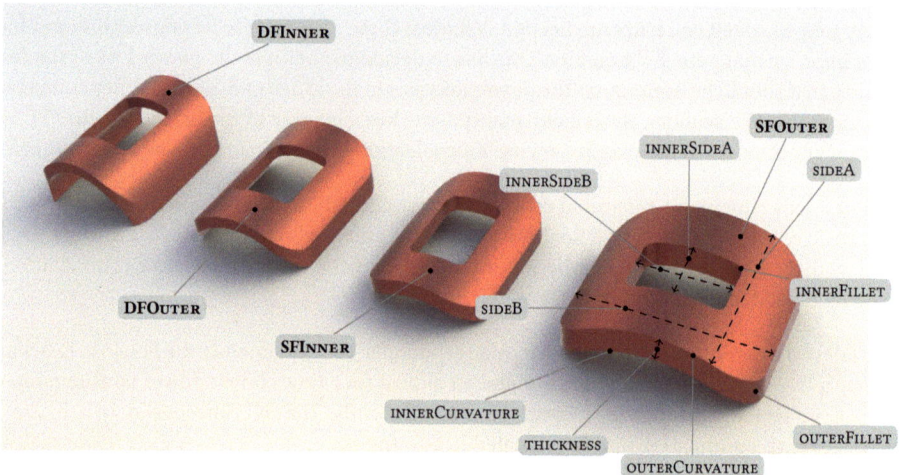

Figure 3.2: Visualization of the four different kinds of coils: inner drive field, outer drive field, inner selection field, and outer selection field coils (from left to right). Geometrical values are illustrated on basis of the selection field coil.

3.2 Coil Controlling

In this section, a brief presentation of the coil controlling and how the currents for the selection field coils have to be chosen is discussed. A sinusoidal function determining the correct currents for every coil array is defined

$$\mathbf{I}_l = \mathbf{I}_l^{\max} \cdot \sin\left(\phi + \frac{\pi}{90} \cdot \gamma\right), \tag{3.1}$$

The current $\mathbf{I}_l^{\max}$ denotes the maximum value for every coil array that is needed to realize an FFL. The variable l indicates one selection field array and γ the angle of the FFL. The phase shift ϕ is unique for every selection field coil array (see table 3.2). Figure 3.3 shows a plot

	SFI1 & SFI3	SFI2 & SFI4	SFO1 & SFO3	SFO2 & SFO4
ϕ	$\pi/2$	$-\pi/2$	0	π

Table 3.2: Corresponding phase shift ϕ for the selection field coils. Figure 3.3 illustrates this for the appropriate angle γ of the FFL.

of equation (3.1). Here, the four different current trajectories are plotted for the appropriate selection field coil array. The rotated FFL is visualized above and below the axis. It should be noted that the amplitudes of the current trajectories are normalized in this plot. The amplitude of the outer selection field ring is usually much higher than the amplitude of the inner selection field ring.

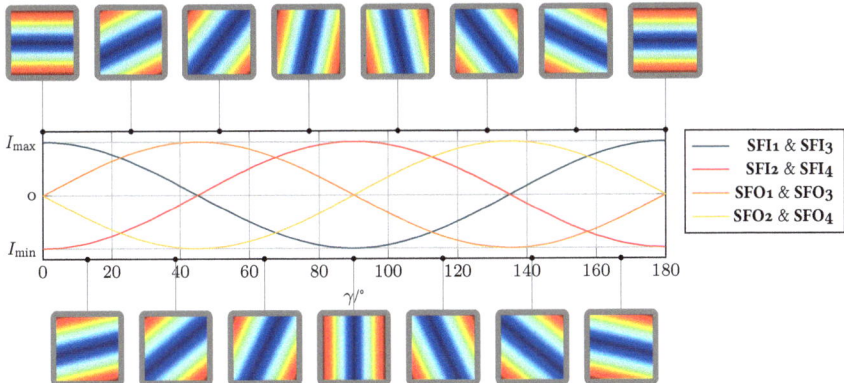

Figure 3.3: The selection field arrays are driven by a current trajectory with a certain phase shift ϕ. This plot exemplary shows these functions. The FFL is visualized below and above the axis for discrete angles.

Finally, the direction of electric current is investigated. Since opposing selection field coil arrays shall generate a gradient field, the direction of electric current has to agree with this fact. Hence, opposing arrays are connected in series and feature diverse winding direction.

3.3 Coil Construction

In section 3.1 the different kinds of coils including geometric properties and function in the complete coil design were discussed. Since the proposed scanner uses a complicated coil geometry, an access to conventional and purchasable coils is not possible. The next section summarizes the coil manufacturing process and its requirements.

3.3.1 Coil Forms

The basic parameters for coil production are provided by the simulation. Coil size and geometry are given and need to be transferred into coil forms. In this work a material combination for the coil forms is used. The inner part of the form with direct contact to the coil is made of polyoxymethylene (POM). Since epoxy resin is used during the winding process of the coils, POM is a useful material choice. Given that POM has a very low surface energy the epoxy resin does not stick to it and allows a simple and fast cleaning process. Further properties are a high strength, hardness and stiffness.

The outer part of the form encloses the POM construction. Here, Aluminum is used. It offers a sufficient physical stability and a good thermal conductivity being important during the com-

3 Materials and Methods

pression molding and heating process. The four different forms are visualized in figure 3.4.

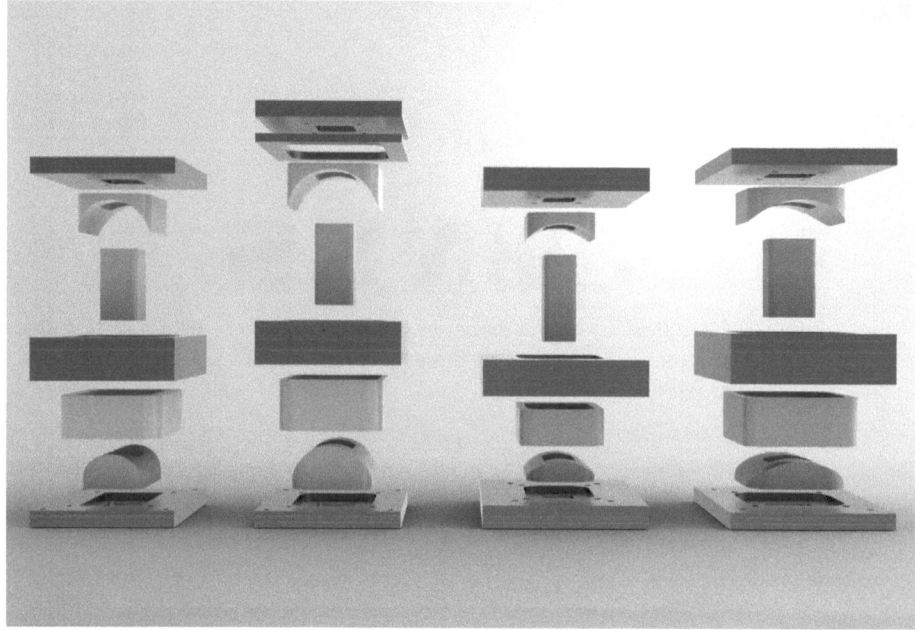

Figure 3.4: Four different forms are constructed. These forms provide proper construction of inner and outer drive field coil respectively inner and outer selection field coil (from left to right).

3.3.2 Litz Wire

The curved coil design has certain requirements on the wire being used for the coil construction. Here, the choice is to use two different kinds of litz wire: a thin variant with a diameter of 2.25 mm and a slightly thicker wire with 3.00 mm diameter. Litz wire consists of thousands of single wires that have a coat of lacquer. Further parameters are listed in table 3.3. The flexibility

	Small litz wire	Big litz wire	Bare wire
Diameter in mm	2.25	3.00	$5.00 \cdot 10^{-2}$
Amount bare wire	1,000.00	2,000.00	1.00
resistance in $\Omega\,\mathrm{m}^{-1}$	$8.71 \cdot 10^{-3}$	$4.35 \cdot 10^{-3}$	8.71

Table 3.3: The two different kinds of litz wire being used. The right column comprehends the properties of a single bare wire.

3.3 Coil Construction

is an essential condition for the coils (see figure 3.6), since coil edges feature a small radius. If the constructed coils do not agree with the simulated shape, proper field generation cannot be realized.

Besides this property, litz wire features a second characteristic. It is especially utilized when it comes to applications requiring alternating currents with high frequencies. These frequencies lead to the skin effect resulting in sharp increases of the resistance respectively power loss [56]. The skin effect is characterized in the next section.

Skin Effect

The time-related alternating current in a conductor induces a time-related magnetic field being caused by Ampères circuit law (see equation (2.36)). On the other hand, this magnetic field induces an eddy current (see equation (2.35), Faraday's law of induction) flowing contrariwise to the applied current. Figure 3.5 visualizes that a higher countervoltage is induced in the center of the wire since there is the highest magnetic field. By increasing the frequency the induced current almost cancels out the applied current. Thus, the currents bulk flows on the conductors surface – this is called skin effect. Resulting from this effect the active cross section decreases and correspondingly the resistance increases. This is critical for the power loss. In equation (2.3), a linear relation concerning the resistance R is derived. By expanding the named equation with respect to the cross section of the conductor

$$P = \rho \cdot \frac{l}{F} \cdot I^2, \tag{3.2}$$

the power loss increases inversely proportional to the cross section F. Since power loss minimization is aimed, this effect has to be avoided. The current density is distributed in the conductor as follows

$$j = j_{\text{surface}} e^{-\frac{d}{\delta}} \quad \text{with} \quad \delta = \sqrt{\frac{\rho}{\pi f \mu}}. \tag{3.3}$$

The variable d characterizes the distance from the surface and δ is a material specific and frequency dependent function. The skin depth δ can easily be calculated for the copper being used in this work

$$\delta = \sqrt{\frac{8.706\,\Omega\,\text{m}^{-1} \cdot 0.001{,}963\,\text{mm}^2}{\pi \cdot 25{,}000\,\text{Hz} \cdot 4\pi \cdot 10^{-7}\,\text{H}\,\text{m}^{-1} \cdot 0.999{,}993{,}6\,\text{H}\,\text{m}^{-1}}} \tag{3.4}$$

$$= 0.4161\,\text{mm}. \tag{3.5}$$

The skin depth δ defines the current density drop by a factor of e^{-1}. Here, it amounts $\delta = 0.4161$ mm and would be critical for the coils being driven by 25 kHz. This effect can be neglected since litz wire with a bare diameter of 0.05 mm is used.

3 Materials and Methods

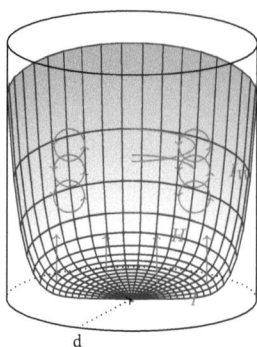

Figure 3.5: A cross section of a conductor is visualized. The applied alternating current I is directed upward and generates a magnetic field H. On the other hand, the magnetic field induces an eddy current being opposite to the applied current. In the background, the current density distribution is plotted as function of the distance d to the surface. With increasing frequency the current density increases near the surface with an exponential distribution.

3.3.3 Winding Process

The coils are manually winded. Every coil has to pass through a specific work chain. This chain is summarized below:

- Coil form cleaning with ultrasonic bath
- First functional solder joint – input/output
- Preparation of epoxy adhesive
- Coil winding – use of thread simplifies winding process
- Second functional solder joint – output/input
- Compression moulding with coil heating/curing

	DFInner	DFOuter	SFInner	SFOuter
Windings	5	5	6	6
Layers	2	2	4	4
Parallel winding	–	–	✓	✓
Small litz wire	–	–	✓	–
Big litz wire	✓	✓	–	✓

Table 3.4: Important attributes for the coil winding process for drive and selection field coils.

The first task is to clean the coil form. Any contamination with magnetic particles could add signal and interfere with the signal quality. Afterwards, the correct litz wire for the appropriate coil has to be chosen. A functional solder joint determines the quality of the coil. Bad solder joints result in heat peaks and a higher resistance that downgrades the power loss.

The next part is the actual winding process. The correct winding direction, amount of windings/layers, and parallel winding has to be considered (see table 3.4). The use of thread can simplify the procedure (see figure 3.6). A proper amount of epoxy resin has to be added to every winding. UHU PLUS 300 KG is a temperature stable high-strength two-component epoxy adhesive and used is in this work.

Finally, a second solder joint finishes the coil. The following compression moulding and heating/curing guarantees the final coil shape. This process utilizes approximately one hour. Overall the coil manufacturing can take up to three hours.

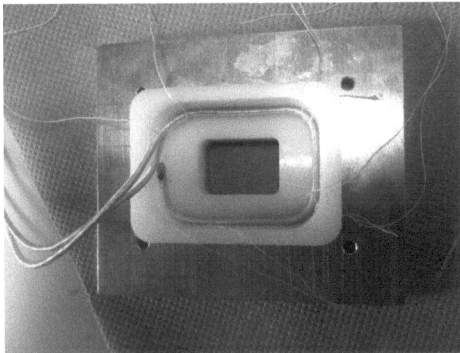

Figure 3.6: The picture demonstrates, how the litz wire is winded into the coil form. Threads are used to strengthen the construction. The litz wire has to be flexible since the edges of the form feature a small radius.

3.4 Coil Assembling

This section describes the strategy for the coil assembling. Certain functionalities have to be kept in mind. Since the design includes 24 coils being aligned around a bore, specific parameters such as

- consistent coil distances,
- equidistant allocation on a circle and
- proper adjustment

3 Materials and Methods

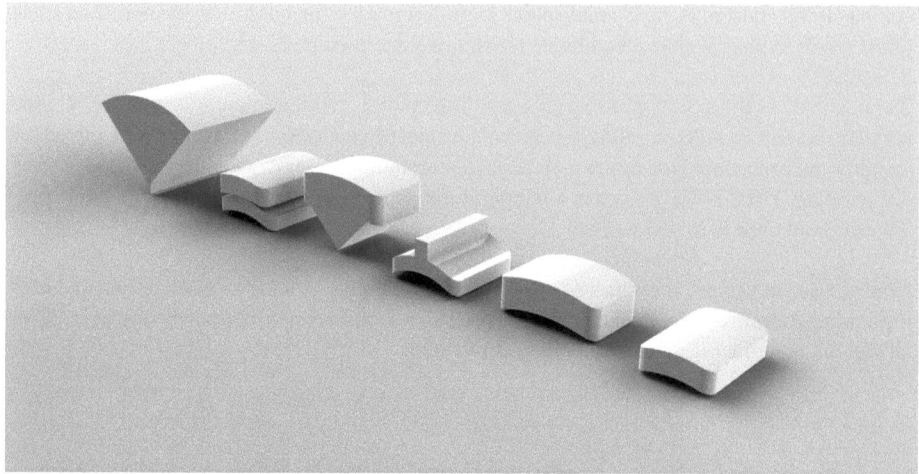

Figure 3.7: Six different plugs and spacer are designed to manage an accurate coil assembling.

have to be maintained. A violation of these parameters could result in a quality loss of the FFL and moreover a higher electrical power consumption. This means e.g. an inhomogeneous gradient field, inconsistent FFL rotation respectively shifting, which would produce image artifacts and worsen the resolution.

A second requirement is to implement an air cooling system. Air shall flow through the coils and is necessary to manage the power loss. The coil configuration features slits between the single coils as well as some bigger gaps between the selection field coil arrays (see figure 3.1). Since an airstream follows the path of minimum resistance, the main flow would directly go through these gaps and a reduced coil cooling would be the result. Hence these gaps have to be closed.

The idea is to combine stabilizing and gap closing parts. Figure 3.7 visualizes these six parts. The following describes how these parts are used and how the coils are assembled (see figure 3.8):

a) The inner drive field coils are assembled on a tubing (not shown for better visualization) and the inner drive field plugs/spacer are installed,

b) the outer drive field coils are perpendicular arranged – the spacers guarantee correct alignment,

c) the outer drive field plugs/spacer are installed,

d) inner selection field array joining is easily possible,

e) the remaining gaps of the selection field coils are closed with plugs,

f) outer selection field plugs/spacer fill the clearances,

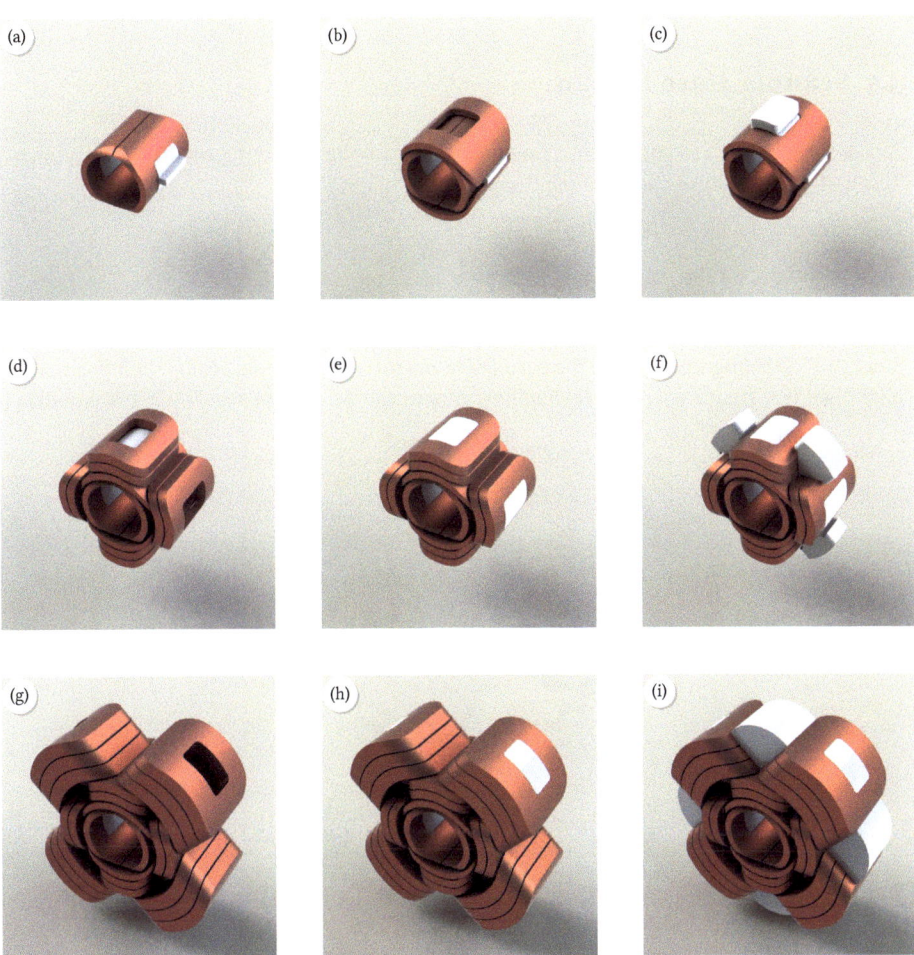

Figure 3.8: This visualization shows, how the coil configuration shall be assembled. The white parts (spacer and plugs) are used to fill gaps and avoid air turbulences. Furthermore, they help aligning the coil arrays. A perpendicular alignment of inner and outer drive field coils is achieved.

3 Materials and Methods

g) the outer selection field coils can easily be fixated,

h) the gaps of the outer selection field coils are closed with the, corresponding plugs

i) the remaining space is filled.

3.5 Scanner Case Design

A task of this work is to implement a customized case for the coil configuration from section 3.1. This case has to feature mainly three components:

a) Cooling device for the coils,

b) an holder for the coil configuration and

c) an attachment for the permanent magnets.

Since the coil configuration with the permanent magnets is fixed concerning their position and alignment (see figure 3.9), only the cooling system can be customized. Thus, the case has to

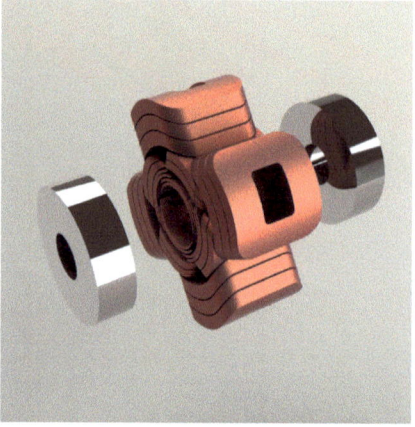

Figure 3.9: Coil configuration with permanent magnets aligned on the z-axis. This setup cannot be changed and the case has to be customized.

be designed around the field generating components. The main idea is to use a design that is easy to assemble and whereat parts – like e.g. permanent magnets or even broken coils – can easily be replaced. The coils should be in a closed compartment with an air in- and output. The bore has to facilitate inserting a Hall probe, phantoms, receive coils, and a mouse for imaging purposes.

Basically, the case consists of four main plates being parallel aligned. The two outer ones fix the permanent magnets, the coil configuration is placed between the inner ones. The outer plates and the permanent magnets itself are described in the following.

3.5 Scanner Case Design

Figure 3.10: Magnet plate of the scanner. The permanent magnet has a copper shielding and is mounted at the center. It is surrounded by six holes enabling proper air supply for the fans. Four small holes connect the different plates with threaded rods.

3.5.1 Permanent Magnet Plates

The permanent magnets are shielded within a copper case to prevent any influences from alternating fields that could change their magnetization. These can be inserted in the POM plate. The six big holes allow air supply for the fans. Threaded rods connect and stabilize the scanner. Two magnet plates shall be realized to generate the gradient for the selection field. The plate is visualized in figure 3.10. The second plate does not feature the air supply holes.

(a) Front view of the system.   (b) Back view of the system.   (c) Used radial fan RL 48-19/12 from EBMPAPST.

Figure 3.11: Air cooling system for the scanner. Six fans are aligned on the ring-shaped construction to provide the air stream for the coils. The air is rerouted with a cylindrical cone.

3 Materials and Methods

Figure 3.12: Acrylic glass that connects air inlet and outlet. It seals the coil configuration airtight and features air hoses at its end.

3.5.2 Air Cooling System

The air cooling system is installed on a further plate. In addition to that, this plate has to fixate the coil configuration. Here, radial fans are used since space is limited between the magnet plate and the cooling system. Moreover, the bore has to be accessible. The concept is visualized in figure 3.11.

Figure 3.13: Air outlet system being realized with a third plate. Air slits provide air removal and a cable lead-through is realized.

The air cooling system is installed on a ring being fixed on POM plate. Since the coil configuration has to be cooled with an air stream flowing in z-direction. Hence, a further part has to reroute the air stream. A cylindrical cone is inserted in the fan array that has the form of a tube and acquires this task.

The coil cooling is realized with common desktop computer fans. So the model RL 48-19/12 from EBMPAPST (see figure 3.11c) is used.

3.5.3 Coil Configuration Sealing Part

It is clear that the coil configuration is located at the center of the scanner – between air inlet and outlet. This structure has to be closed by a further part. An acrylic glass sealing the coils air-tight serves this purpose. Furthermore, it allows a view inside the construction. The acrylic glass is shown in figure 3.12. It has a circular shape and features air holes facilitating the air outlet.

3.5.4 Air Outlet System

After the air stream cools the coil configuration and is heated up, it has to be released. Thus, the third scanner plate features an air outlet system. Circular aligned air slits enables air removal. Below the hole for the bore, a second opening facilitates cable lead-through.

3.5.5 Scanner Assembling

After introducing the individual parts, an overview of the complete scanner construction shall be given. Figure 3.14 visualizes this process. First, the permanent magnet plate and the air cooling system are assembled with the help of the four threaded rods (a). Then, the acrylic glass, the inner tube and the coil configuration are installed (b). The air outlet system finalizes the cooling system and seals the coils air-tight. Finally, the second magnet plate completes the scanner case (d).

3.6 Magnetic Field Measurements

The evaluation of the constructed coil configuration and the simulated results from [8] is crucial. To get first results, a simplified permanent magnet and coil holder (see figure 3.15) is built. In the following the realization of the measurements is described.

3 Materials and Methods

Figure 3.14: This visualization shows how the scanner shall be assembled. First, the magnet plate and the cooling plate are assembled with four threaded rods. Afterwards, the coil configuration is mounted on the bore and the acrylic glass seals the cooling system. Finally, the air outlet and the second magnet plate complete the scanner case.

3.6.1 Measurement Environment

The measurements are done with a three-dimensional Hall Probe (LAKESHORE, see figure 3.15a and figure 3.15b on the right hand side). The Hall probe is mounted on a robot (ISELAUTOMATION GMBH & Co. KG) and can be controlled by a computer. However, the measured magnetic field

3.6 Magnetic Field Measurements

(a) Side view.

(b) Top view.

Figure 3.15: A simple construction that holds the permanent magnets and the coil configuration. Two fans are used to cool the coils. On the right the Hall probe can be seen.

data has to be modified, since the three sensors for the corresponding field direction do not feature the same position. Figure 3.16 demonstrates this for the x- and y-sensors. Unlike the z-sensor, these sensors are shifted by 2.08 mm in perpendicular directions from the geometric center of the probe. Thus, one has to compute an interpolation to gain the correct magnetic field values and positions.

The coils are driven by six current sources: four current sources (**DELTAELEKTRONIKA** SM 7.5-80) for the selection field coils and two (**SORENSEN** DLM 8-75) for the drive field coils. The current is manually set with a precision of one significant figure after the decimal point.

3.6.2 Parameter Setup

A rotated respectively shifted FFL needs proper coil controlling such as choosing the correct currents and a permanent magnet adjustment for a desired gradient. With the simplified setup a gradient of $0.4\,\mathrm{T\,m^{-1}}\,\mu_0^{-1}$ shall be realized. Hence, all required parameters have to be simulated and evaluated respectively adapted afterwards. Hereafter the permanent magnet adjustment is described, followed by the measurement process, and data acquisition.

Permanent Magnet Adjustment

The permanent magnets in this FFL scanner concept generate the necessary gradient. On the one hand, the gradient is dependent on the distance between both magnets and on the other hand, on the material specific remanence. With given remanence, the distance can be calculated and validated. Since the used NdFeB permanent magnets are only labeled with a certain range about approximately 1 T, another approach has to be considered. A time consuming

3 Materials and Methods

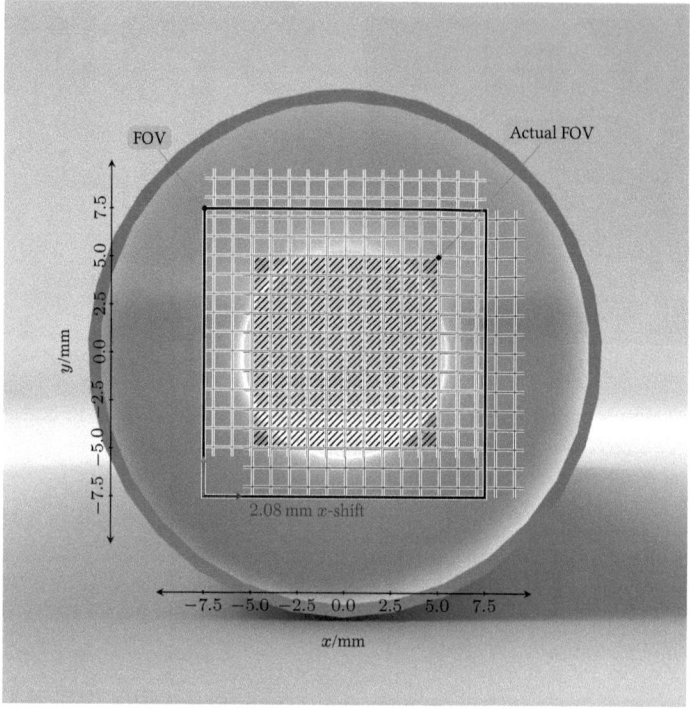

Figure 3.16: Offset of the measured magnetic field components. The x- and y-sensors of the Hall probe are perpendicular aligned, but shifted about 2.08 mm compared to the geometric center of the probe. Thus, the actual magnetic field has to be interpolated and the actual FOV is smaller (highlighted area). The background shows the bore that limits the FOV.

method would be an iterative process of adjusting the distance and measuring the field. Here, an alternative method is proposed.

First, the magnetic field between both magnets at a certain distance is measured. With given size and shape of the magnets the remanence can be recalculated

$$0 = \mathbf{H}_i^{\mathrm{meas}_z}(\mathbf{r}_i) - \mathbf{H}_i^{\mathrm{sim}_z}(\mathbf{r}_i, B_{\mathrm{remanence}}) \quad (3.6)$$

by solving this equation numerically. Therefore, a Gauss-Newton-Solver is implemented. With the calculated respectively measured remanence the distance for desired gradient can be computed with a similar approach

$$0 = \mathbf{G}^{\mathrm{dest}_z} - \mathbf{G}^{\mathrm{sim}_z}(d_{\mathrm{magnets}}). \quad (3.7)$$

3.6 Magnetic Field Measurements

This setup consists of two permanent magnets on each side. They are separated by a wooden plate. Their geometric parameters are listed in section 3.6.2.

	Value in m
Diameter	$8.50 \cdot 10^{-2}$
Thickness	$2.85 \cdot 10^{-2}$
Length	$8.00 \cdot 10^{-3}$

Table 3.5: Geometric parameters of the permanent magnets that are used.

Measurement Procedure

Magnetic field measurements are generally sensible processes being very interference-prone for certain influences. These influences are explained in the following. The geometrically correct alignment is critical. On the one hand, the coil configuration and the permanent magnets have to be adequately assembled. On the other hand, these components have to be justified with the Hall probe, the robot arm, and the robot itself. Thus, many errors can occur on this geometric chain.

Further sources of errors can be of electronically nature. The Hall probe has a limited accuracy and can be influenced by interfering fields such as the static magnetic field of the earth. This can be bypassed by calibrating the Hall probe. Further magnetic field influences are avoided by relocating the measurements in a shielded room.

Last, since the Hall Probe is discretely moved over the FOV, start and stop processes of the probe result in a vibration. By giving the hall probe enough time for relaxation, this influence is minimized.

Now, the geometrically correct alignment is reconsidered. A measurement of 25×25 pixels takes up to 20 minutes. An incorrect alignment results in a re-justification and a repetition of the measurement process. Thus, an optimized algorithm is implemented solving the alignment process as well as the temporal component. The idea is to do a visual justification of the system and a magnetic field measurement afterwards. This first result is processed by the algorithm and compared to the simulated field. A simple registration process being based on a Gauss-Newton solver calculates the displacement between both fields in terms of translation and rotation. Afterwards, one can use the results to recalculate the center of the FOV and reposition the robot. This method is way more accurate than performing a manual adjustment.

Furthermore, a factor $\mathbf{k}_h$ is calculated that minimizes the error between simulated and mea-

3 Materials and Methods

sured field

$$\mathbf{k}_h = \frac{\sum_{i=1}^{M} \sum_{j=1}^{N} \left(\mathbf{H}_{i,j}^{\text{sim}_h^x}(\mathbf{r}_{i,j}, \mathbf{I}_h) \cdot \mathbf{H}_{i,j}^{\text{meas}_h^x}(\mathbf{r}_{i,j}) \right.}{\sum_{i=1}^{M} \sum_{j=1}^{N} \left(\left| \mathbf{H}_{i,j}^{\text{sim}_h^x}(\mathbf{r}_{i,j}, \mathbf{I}_h) \right|^2 \right.} \dots \quad (3.8)$$

$$\dots \frac{+ \mathbf{H}_{i,j}^{\text{sim}_h^y}(\mathbf{r}_{i,j}, \mathbf{I}_h) \cdot \mathbf{H}_{i,j}^{\text{meas}_h^y}(\mathbf{r}_{i,j})}{+ \left| \mathbf{H}_{i,j}^{\text{sim}_h^y}(\mathbf{r}_{i,j}, \mathbf{I}_h) \right|^2} \dots \quad (3.9)$$

$$\dots \frac{+ \mathbf{H}_{i,j}^{\text{sim}_h^z}(\mathbf{r}_{i,j}, \mathbf{I}_h) \cdot \mathbf{H}_{i,j}^{\text{meas}_h^z}(\mathbf{r}_{i,j}) \Big)}{+ \left| \mathbf{H}_{i,j}^{\text{sim}_h^z}(\mathbf{r}_{i,j}, \mathbf{I}_h) \right|^2 \Big)}. \quad (3.10)$$

This factor $\mathbf{k}_h$ corrects the currents so that the simulation matches reality. The symbol

$$\mathbf{H}_{i,j}^{\text{sim}_h^{x,y,z}}(\mathbf{r}_{i,j}, \mathbf{I}_h) \quad (3.11)$$

is the x-, y- or z-component of the simulated magnetic field of coil array h with respect to the current $\mathbf{I}_h$. Here, the symbol $h \in 1, \dots, 6$ indexes the corresponding magnetic field of the particular coil array: two inner and two outer selection field arrays and one inner and one outer drive field array. The factor $\mathbf{k}_h$ is obtained by minimizing the cost function

$$\mathbf{f}_h(\mathbf{k}_h) = \sum_{i=1}^{M} \sum_{j=1}^{N} \left| \mathbf{H}_{i,j}^{\text{sim}_h^{x,y,z}}(\mathbf{r}_{i,j}, \mathbf{I}_h) - \mathbf{k}_h \mathbf{H}_{i,j}^{\text{meas}_h^{x,y,z}}(\mathbf{r}_{i,j}) \right|. \quad (3.12)$$

This least squares approach allows for calculating the optimized and correct currents $\mathbf{I}_h^{\text{opt}}$

$$\mathbf{I}_h^{\text{opt}} = \mathbf{k}_h \cdot \mathbf{I}_h. \quad (3.13)$$

Analogous to [6], the normalized root mean square deviation (NRMSD) is calculated to compare measured fields with simulated fields [57]. The NRMSD can be calculated considering the root-mean-square deviation (RMSD) and the measured/simulated magnetic field $\left\| \mathbf{H}^{\text{meas,sim}} \right\|_2$

$$\text{RMSD} = \sqrt{\frac{\sum_{i=1}^{n} \left(\left\| \mathbf{H}_i^{\text{meas}} \right\|_2 - \left\| \mathbf{H}_i^{\text{sim}} \right\|_2 \right)}{n}}. \quad (3.14)$$

Thus, the NRMSD is the RMSD divided by the range of measured data

$$\text{NRMSD} = \frac{\text{RMSD}}{\left\| \mathbf{H}^{\text{meas,sim}} \right\|_2^{\max} - \left\| \mathbf{H}^{\text{meas,sim}} \right\|_2^{\min}}. \quad (3.15)$$

Chapter 4

Results

The following chapter presents the constructed curved rectangular coils and its assembling to the coil configuration. Furthermore, the generated respectively simulated magnetic fields are measured and visualized. Thereby the power loss is measured and analyzed. Finally, the complete scanner setup is assembled by mounting coil configuration and permanent magnets to the scanner case.

4 Results

(a) Side view.   (b) Bottom view.

Figure 4.1: An outer selection field coil serves as an example for the coil construction. Both solder joints are located close to each other on one side. The coil features double winding and is built with thread to optimize the winding quality.

4.1 Coil Assembling

The manufactured coils have to be assembled to the FFL generating structure. At first, a single constructed coil is presented. An outer selection field coil serves as an example and is shown in figure 4.1. Here, the winding technique and the threads optimizing the winding quality can be seen. The coil has six windings and four layers.

The whole coil configuration is assembled as described in section 3.4, figure 3.8. The single coil arrays are separated by small nylon spacers with a thickness of 1.5 mm and adhered with epoxy composite. These spacers can be seen in figure 4.2 part (a) and (b). Polyimide film in the form of Kapton is placed between the coil layers to prevent voltage flashovers and enameled silver wire connects the single coils. Moreover, coil connections between opposing arrays are realized with litz wire being also isolated by shrink tubing. The yellow cable connects coils in series and the blue respectively white ones denote junctions to the current sources. Wires leading to the power sources are isolated with black shrink tubing. The wires are circular aligned and feature a small distance to the coils so that a minimum space is required. These single assembling steps are illustrated in figure 4.2. Part (h) displays the final setup including all spacers and plugs.

It can be noted that the blue adhesive does not stabilize the final construction in terms of steadiness. Rather, it simplifies the assembling process. Final stability is provided by installing cable tie around the outer coil ring. Thus, it is possible to quickly disassemble the structure to replace coils or fix solder joints. This option could be used to replace one outer selection field coil showing a high resistance respectively broken litz wire.

4.1 Coil Assembling

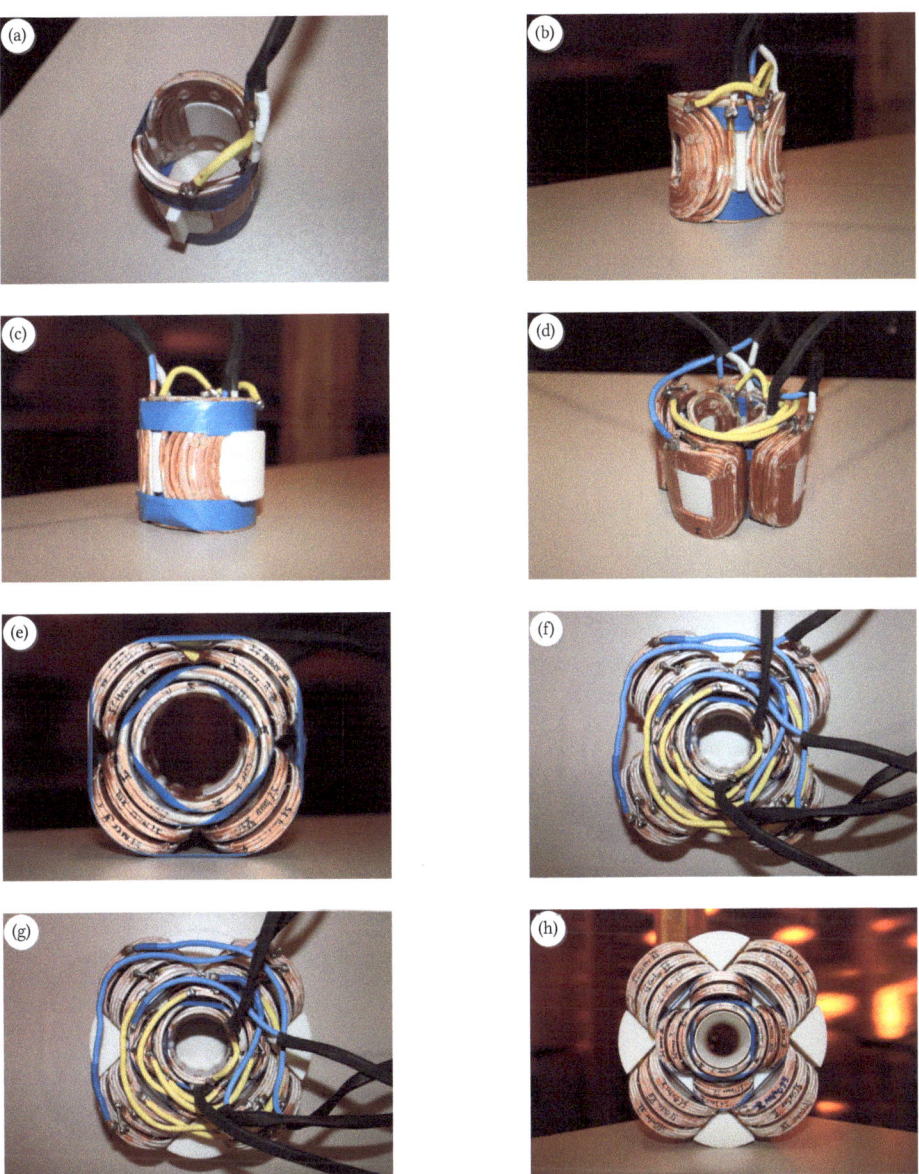

Figure 4.2: This sequence of images shows how the coils are assembled. This is done analogous to section 3.4 respectively figure 3.8. As one can see the coils are already connected and soldered. The interconnecting cables are isolated with shrink tubing.

4 Results

4.2 Field Measurements

4.2.1 Permanent Magnet Adjustment

As described in section 3.6.2, a certain algorithm has to be executed to adjust the permanent magnets. The first step is to measure the z-component of the magnetic field on the center axis between the magnets. Thus, the construction is justified and a distance of 20.7 cm is set between the inner surfaces of the magnets. This distance is simulated by assuming a gradient of $0.4\,\mathrm{T\,m^{-1}}\,\mu_0^{-1}$ and a remanence of 1 T. The measured field is plotted in figure 4.3. It can

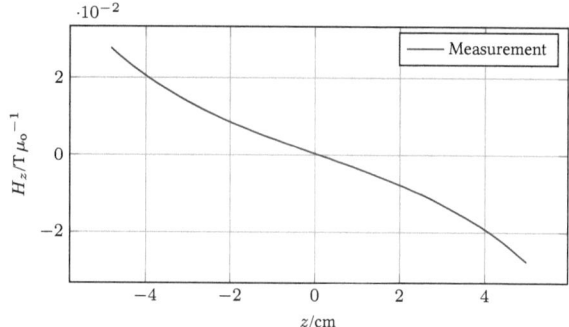

Figure 4.3: Magnetic field measurement between two opposing permanent magnet arrays with a distance of 20.7 cm. A gradient of $3.86\,\mathrm{T\,m^{-1}}\,\mu_0^{-1}$ is achieved.

be noted that a gradient strength of $0.386\,\mathrm{T\,m^{-1}}\,\mu_0^{-1}$ is already achieved with this alignment, which is close to the desired one. The corresponding gradient field is plotted in figure 4.4.

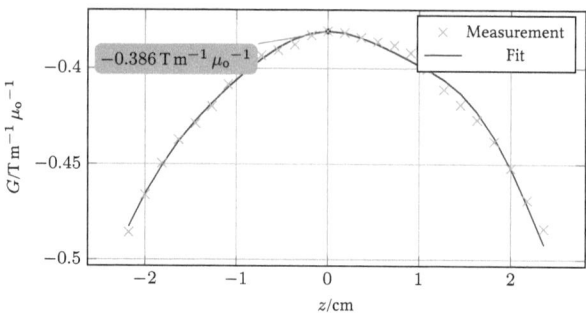

Figure 4.4: Gradient of the magnetic field measurement of figure 4.3. The data is fit with a spline. A gradient of $0.386\,\mathrm{T\,m^{-1}}\,\mu_0^{-1}$ is measured at the center.

Now, the data is used to calculate the real remanence by solving the problem of equation (3.6). Consequently, a remanence of 0.986 T is obtained. This result is used for the next step. By means of the new remanence, equation (3.7) computes a distance of 20.4 cm for a gradient of $0.4\,\mathrm{T\,m^{-1}}\,\mu_0^{-1}$. After re-calibrating the assembly the magnetic field is measured again. The course of the gradient is plotted in figure 4.5. Now, the gradient amounts to $0.411\,\mathrm{T\,m^{-1}}\,\mu_0^{-1}$ and therewith, is slightly higher than the computed value. A further re-adjustment is not necessary and not performed since a distance variation of 3 mm changes the gradient by $0.03\,\mathrm{T\,m^{-1}}\,\mu_0^{-1}$. The simple construction fixating the permanent magnets does not allow justifications that are more precise than 1 mm. This is due to the fact that the wooden permanent magnet holders are slightly flexible.

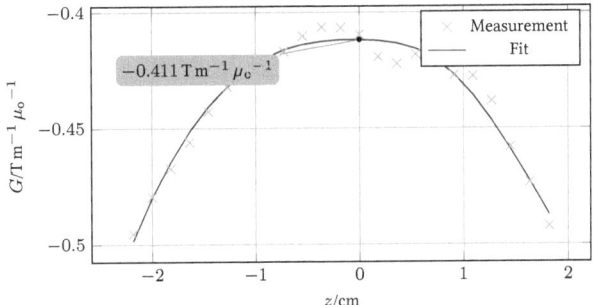

Figure 4.5: Gradient of the magnetic field at a distance of 20.3 cm. The data is fit with a spline. A gradient of $0.411\,\mathrm{T\,m^{-1}}\,\mu_0^{-1}$ is measured at the center.

4.2.2 Results for FFL Rotation

The FFL rotation is split up into eight iterations between $0°$ and $167.5°$ with steps of $22.5°$. A FOV of 25×25 pixel is measured, which has a size of $15\,\mathrm{mm} \times 15\,\mathrm{mm}$. Afterwards, the absolute value is calculated with the algorithm described in section 3.6.1. Hence, the resulting size of the magentic field is $10\,\mathrm{mm} \times 10\,\mathrm{mm}$. The applied currents for the particular angle can be found in table 4.1. The absolute values of the magnetic field is plotted in figure 4.7. Moreover, the simulated fields are shown in figure 4.6 for comparison.

Furthermore, the NRMSD between simulation and measurement is calculated. The results are

Rotation $\gamma/°$	SFI $|I|$/A	SFO $|I|$/A
0.0	9.5	0.0
22.5	6.7	15.1
45.0	0.0	21.3
67.5	6.7	15.1
90.0	9.5	0.0
112.5	6.7	15.1
135.0	0.0	21.3
157.5	6.7	15.1

Table 4.1: Applied currents for the FFL rotation.

4 Results

listed in table 4.2 for every angle. The average NRMSD amounts to 2.39 %. The range of the magnetic field strength is between $0\,\mathrm{T}\,\mu_o^{-1}$ and $2.66 \cdot 10^{-3}\,\mathrm{T}\,\mu_o^{-1}$ for the simulation and between $0\,\mathrm{T}\,\mu_o^{-1}$ and $2.77 \cdot 10^{-3}\,\mathrm{T}\,\mu_o^{-1}$ for the measurement.

Rotation $\gamma/°$	NRMSD/%
0.0	1.53
22.5	4.41
45.0	3.53
67.5	2.72
90.0	1.91
112.5	1.86
135.0	2.05
157.5	1.11

Table 4.2: NRMSD of the measured FFL compared to the simulated one.

4.2.3 Results for FFL Translation

The parameters for the translated FFL are similar to the previous section. A FOV of 10 mm × 10 mm is realized with a discretization of 25 × 25 pixels. Three measurements are performed by realizing three different rotations respectively translations. A shift of 2.5 mm is performed whereas the rotation amounts to 0°, 45° and 135°. The translation is performed perpendicular to the FFL course with the currents listed in table 4.3. The NRMSD between simulated and

Rotation $\gamma/°$	Shift x/mm	Shift y/mm	DFI $|I|$/A	DFO $|I|$/A	SFI $|I|$/A	SFO $|I|$/A
0	0.0	2.5	0.0	4.4	9.5	0.0
45	−2.5	2.5	2.9	4.4	0.0	21.3
90	−2.5	0.0	2.9	0.0	9.5	0.0

Table 4.3: Applied currents for the FFL rotation and translation.

measured magnetic fields is calculated. An average NRMSD of 2.78 % is achieved. The results are also listed in table 4.4. The range of the magnetic field strength is between $0\,\mathrm{T}\,\mu_o^{-1}$ and $3.87 \cdot 10^{-3}\,\mathrm{T}\,\mu_o^{-1}$ for the simulation and between $0\,\mathrm{T}\,\mu_o^{-1}$ and $3.73 \cdot 10^{-3}\,\mathrm{T}\,\mu_o^{-1}$ for the measurement.

Rotation $\gamma/°$	NRMSD/%
0.0	4.42
135.0	1.50
90.0	2.42

Table 4.4: NRMSD of the measured FFL field compared to simulated data.

4.2 Field Measurements

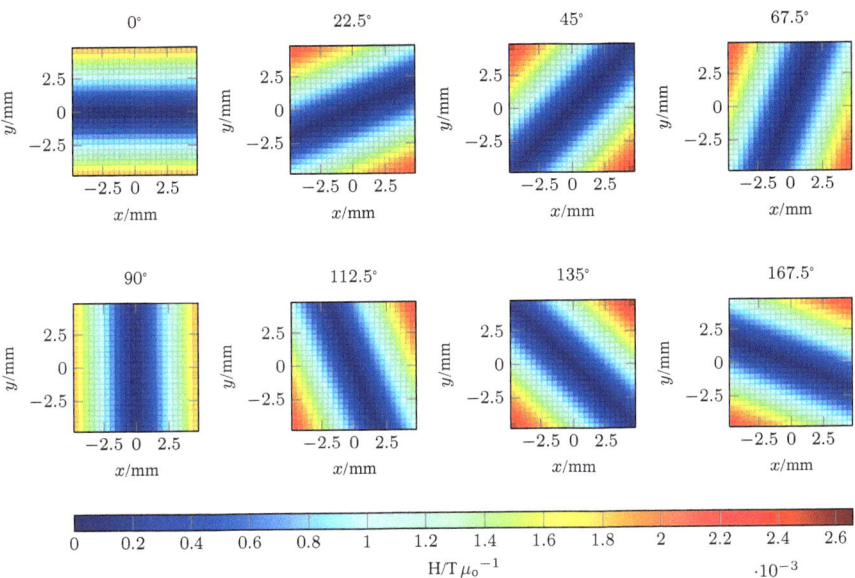

Figure 4.6: Absolute value of the simulated FFL fields. The FFL is rotated stepwise by 22.5° from 0° to 167.5°.

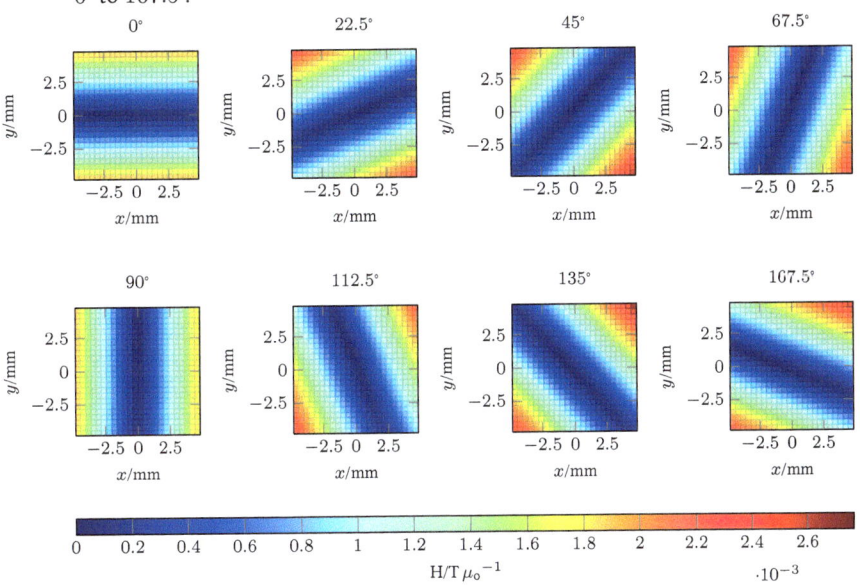

Figure 4.7: Absolute value of the measured FFL fields. The FFL is rotated stepwise by 22.5° from 0° to 167.5°.

4 Results

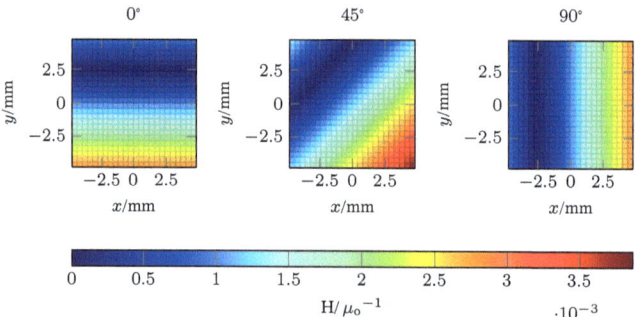

Figure 4.8: Magnetic field simulations of a rotated and shifted FFL. The FFL is rotated by 0°, 45° and 90° and shifted by 2.5 mm in the corresponding direction.

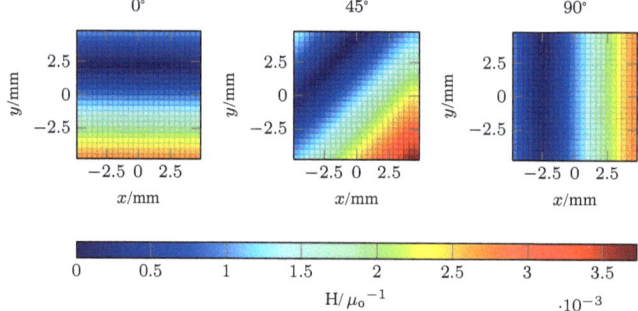

Figure 4.9: Magnetic field measurements of a rotated and shifted FFL. The FFL is rotated by 0°, 45° and 90° and shifted by 2.5 mm in the corresponding direction.

4.2.4 Power Loss

The power loss is calculated for the FFL rotation and translation. The dc current sources are set with the maximum currents and subsequently the voltage is tapped at every coil package. With the measured voltage and the applied currents, equation (2.11) allows for calculating the power loss. The measured results and the calculated power loss are listed in table 4.5. In addition, the simulated power loss is displayed. The average measured power loss for inner drive field coils $P^{DFI}_{mean} = 0.0625\,W$, outer drive field coils $P^{DFO}_{mean} = 0.1545\,W$ and inner selection field coils $P^{SFI}_{mean} = 1.6013\,W$ is in the range of the simulated one. It can be noted that the first package of the inner selection field coils has a slightly higher power loss than the simulation. The outer drive field coils power loss amounts to $P^{SFO}_{mean} = 7.7378\,W$ and is also slightly higher than the simulated power loss. In summary the measured power loss is 37.9 W and the simulated one amounts to 32.5 W.

		Ampère in I/A	Voltage in U/V	Measured power loss in P/W	Simulated power loss in P/W
DFI	1	2.9	0.021	0.062	0.065
	2	2.9	0.022	0.063	0.065
DFO	1	4.4	0.035	0.155	0.149
	2	4.4	0.035	0.154	0.149
SFI	1	9.5	0.196	1.859	1.590
	2	9.5	0.162	1.534	1.590
	3	9.5	0.160	1.517	1.590
	4	9.5	0.157	1.495	1.590
SFO	1	21.3	0.368	7.838	6.539
	2	21.3	0.352	7.506	6.539
	3	21.3	0.388	8.273	6.539
	4	21.3	0.345	7.334	6.539
Overall				37.799	32.484

Table 4.5: The measured voltage and the applied current allow for calculating the power loss. The overall measured power loss is slightly higher than the simulated one.

4.3 Scanner Case

In the following, the manufactured scanner case is presented. Beginning with the description of single parts, it is approached to the assembling of permanent magnets and coil configuration into the case concept.

4.3.1 Air Cooling System

The air cooling system is the centerpiece of the case since it cools the coils and guarantees its fixation. Figure 4.10 demonstrates one important part of this structure. Due to its complex design it is rapid prototyped. Six circular aligned fan ports are provided serving as air inlet

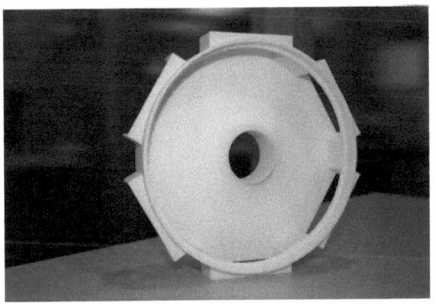

(a) Side view.

(b) Bottom view.

Figure 4.10: The constructed air cooling holder rerouting the air from the radial fans to the coil configuration.

4 Results

Figure 4.11: The constructed acrylic glass featuring air hoses and a rejuvenation so it can be connected to the air cooling system.

(see figure 4.10a). The center hole has a diameter of 38 mm. On figure 4.10b one can see the curved cone redirecting the air stream. Additionally, the system is enclosed by an acrylic glass featuring the necessary air outlets. It can be attached to the air cooling holder and is shown in figure 4.11. It has an outer diameter of 170 mm and an inner diameter of 160 mm. Both parts are locked between two POM plates (see figure 4.13).

4.3.2 Plate Magnet

The permanent magnets are placed into a circular copper box. This box has an outer diameter of 104 mm, an inner diameter of 26 mm and a length of 30 mm. It consists of four parts shown in figure 4.12b. The shielded permanent magnets are fixed in the POM plate displayed in figure 4.12a, which features a small indentation. The POM plate has a size of

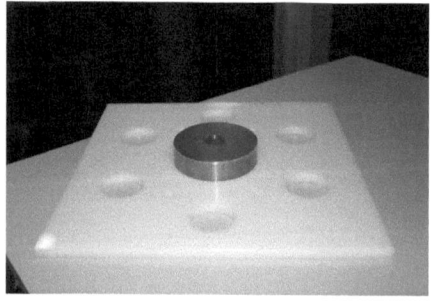
(a) Plate magnet.

(b) Copper box.

Figure 4.12: The shielded permanent magnet is assembled on the magnet plate, which is constructed of POM.

Figure 4.13: Stepwise visualization of the scanner case assembling. The four POM plates fixate the shielded permanent magnets and the coil configuration at the center which is cooled by an air cooling system.

4 Results

300 mm × 300 mm × 10 mm as well as the other three plates (see figure 4.13).

4.3.3 Assembling

The remaining scanner parts and the case assembling is explained in the following (see figure 4.13). First, the completed magnet plate is connected to the threaded rots and fixated to a second plate of the same size (a). Subsequently, the air cooling system from figure 4.10 and figure 4.11 can be installed (b). A tube consisting of fiberglass with an outer diameter of 38 mm and an inner diameter of 36 mm is attached to the bore. In part (c) and (d), the coil configuration is aligned on this tube and stabilized by a third plate featuring the air outlet. This star-shaped air outlet facilitates the externalization of the cable routing (e). Finally, the second magnet plate is attached and completes the case (f). The radial fans can be installed as shown in (g). It can be noted that the air supply for the fans is guaranteed (h).

Chapter 5

Discussion

The realized curved rectangular coils, its assembling, and the generated fields are discussed in the following. Furthermore, a short outlook gives an overview of future work.

5 Discussion

Four different curved rectangular coil forms could be implemented. These forms facilitated the winding of the coils. This process chain is critical, since any mistake results in bad functioning coils. Broken litz wire or poor solder joints result in higher resistances. Higher resistances downgrade the power loss and increases the coils temperature. Thus, one has to carefully consider proper construction. Furthermore, an optimized winding technique was applied that uses thread to stabilize the litz wire during the winding process.

Despite the ambitious coil design, all 24 coils were implemented and show a great quality in terms of visual assessment.

Subsequently, the assembling of the coils was planned. Several plugs and spacers were designed to simplify the construction. Furthermore, these parts had to enhance the air cooling efficiency by closing gaps in the coil array. These parts simplified the assembling to a minimum and resulted in great stability of the whole construction. However, if it comes to broken coils or optimizations in the future, this construction allows easy repair and optimization mechanisms, since one focus was a flexible design. The magnetic field of a rotated and shifted FFL could be realized with a gradient of $0.4\,\mathrm{T\,m^{-1}}\,\mu_o^{-1}$. This was verified by measuring the field with an Hall probe and a robot system. In the following, these steps are briefly summarized.

First, permanent magnets generating the desired gradient had to be adjusted. This adjustment was split up into two parts: the permanent magnets remanence was measured and afterwards, used to align the setup. The implementation of a Gauss-Newton solver could drastically accelerate this task and avoid an iterative process of adjusting and measuring the fields. Finally, a gradient of approximately $0.41\,\mathrm{T\,m^{-1}}\,\mu_o^{-1}$ could be achieved. Further optimizations were not necessary since the construction fixating the permanent magnets did not allow more precise adjustments.

Next, the manufactured coil configuration was added to the permanent magnets and connected to current sources. Pre-simulated currents were applied to generate eight equidistant rotations of the FFL. The average NRMSD amounted to $2.39\,\%$ and thus, is in a range comparable to [8]. An algorithm comparing simulated and measured fields being based on a Gauss-Newton solver facilitated very precise adjustments of the Hall probe. The algorithm performs an registration process between the precalculated magnetic field and the measured two-dimensional field. That simplified the whole measurement process since a manual readjustment was not necessary.

Besides the pure FFL rotation, a further FFL translation was realized. Thereby, the FFL was translated by $2.5\,\mathrm{mm}$ in three different directions for three different rotations. An average NRMSD of $2.78\,\%$ between measured and simulated field was calculated.

Despite the minimal difference between measured and simulated fields, there can exist several error sources. On the one hand, the inaccurate alignment between field generator and Hall probe respectively robot result in displacement between the fields. On the other hand, the current sources were controlled manually. The currents could only be set with an accuracy concerning first decimal place. Furthermore, based on the iterative movement of the Hall probe, the starting and stopping process results in a vibration of the sensor, leading to a displacement concerning the actual position.

However, excellent results were achieved being a good index of accurate coil manufacturing.

In addition to the magnetic field, the power loss was calculated by measuring the voltage at a certain current. The power loss of the inner drive field coils, outer drive field coils and inner selection field coils showed excellent agreement with the simulated one. However, one inner selection field coil and all outer selection field coils featured a slightly higher power loss. This might be a result of broken litz wire inside the coil or bad solder joints. Since three outer selection field coils showed almost the same power loss, there might be a correlation between compression moulding, coil form and broken litz wire. Perhaps litz wire broke consistently at the same location during the compression moulding process as a result of sharp edges or too much pressure. In summary, the measured power loss was 37.9 W and the simulated one 32.5 W. It can be noted that the power loss of this implementation was conspicuously smaller than the FFL demonstrator from [6] which amounts to 42.32 W (only rotation) and generates a 1.64 times smaller gradient. Since gradient and applied currents have a linear relation it is possible to calculate the power loss reduction for the same gradient:

$$37.9\,\text{W} = (1.64 \cdot I_{0.25\,\text{T m}^{-1}\,\mu_0^{-1}})^2 \cdot R \tag{5.1}$$

$$\Rightarrow 14.09\,\text{W} = I^2_{0.25\,\text{T m}^{-1}\,\mu_0^{-1}} \cdot R. \tag{5.2}$$

That results in three times lower power loss at a gradient of $0.25\,\text{T m}^{-1}\,\mu_0^{-1}$. This is a drastic enhancement of power loss optimization.

The design of a scanner case is a challenging work, since a lot of variables and conditions have to be considered. The focus was to install an air cooling system and a fixation for the field generating parts. The basic concept features four POM plates whereas the outer ones fixate the permanent magnets and the inner ones the coil configuration as well as the cooling system. The resulting case was partly rapid prototyped respectively manufactured.

The single parts were easy to assemble leading to the result that designing work and preliminary considerations were well chosen. Furthermore, the whole flexibility future-proofs the scanner case since e.g. permanent magnets can easily be replaced. This is an important point because the permanent magnet shielding limits the bore to 26 mm and therefore, the whole bore diameter of 36 mm is not used. Future works might tap out the full potential by installing permanent magnets with a bigger inner diameter.

An additional idea is to extend the scanners functionality by installing customized z-coils in combination with permanent magnets. That would offer the possibility to upgrade the FOV by a third dimension. Than, full three-dimensional FFL imaging would be possible. So far, this was only possible by moving the object [45]. However, further FFL trajectories have to be analyzed resulting in an expanded Radon-based reconstruction technique.

Nevertheless, the efficiency of the cooling system and its implementation have to be proven.

So far, the focus was to generate an FFL with the coil configuration. The whole construction is motivated by aspects such as higher sensitivity, higher SNR and faster reconstruction methods – all in comparison to conventional FFP imaging. These aspects are only of simulated nature

5 Discussion

and have not been proven experimentally. Hence, this proof is crucial. A further property of the coil configuration is not just the FFL generation, but also an FFP generation. This could be done by removing the permanent magnets and generate the gradient by the selection field coils and move the FFP by the drive fields. Moreover, permanent magnets or mentioned z-coils could generate the FFP and selection and drive field coils would incur the FFP movement. Than, FFP respectively FFL imaging could be realized in one scanner. This offers great possibilities to compare both spatial encoding schemes in terms of sensitivity, resulting SNR and difficulty of image reconstruction.

To complete this scanner topology, further steps have to be taken. Filters for signal generation and reception have to be implemented and image reconstruction and controlling realized. Furthermore, appropriate phantoms that fit in the bore have to be manufactured.

An interesting aspect that is most important for the imaging process, is the influence of the drive field coils concerning the selection field coils and vice versa. The induced voltage can change the quality of the magnetic field. Eventual field inconsistencies have to be analyzed and avoided by certain techniques such as extra shielding. This shielding could be realized between inner selection field ring and outer drive field ring by installing a copper tube.

Signal generation from the fans is most likely. Thus, first scans will proceed with a discrete cooling. FFL movement respectively signal generation is done when the fans are switched off.

Chapter 6

Conclusion

6 Conclusion

The fundamental interrogation of this thesis is the realization of curved rectangular coils optimizing field quality and minimizing the power consumption in an FFL setup. After introducing the problem outline, the physical principles of MPI have been carried out. Subsequently, procedures were listed describing the assembling and analysis of the FFL generating setup. Finally, the presented results were discussed and evaluated.

The presented work shows that an FFL of high quality and a power loss being equal to an FFP scanner with same gradient strength and size can be practically realized. Resulting from this consequence, the sensitivity in MPI can be improved by one order of magnitude without developing expensefull cooling systems such as an oil cooling or water cooling device. Furthermore, the high FFL quality allows reconstruction techniques that does not need the time consuming recording of the system matrix and therefore, simplifies and improves real-time imaging without a loss of image quality.

The setup allows rotation as well as translation of the FFL. An evaluation was done by comparing measured fields to simulated ones resulting in a mean NRMSD below 3%. A gradient of $0.41\,\mathrm{T\,m^{-1}}\,\mu_0^{-1}$ was realized. Besides, the power loss amounted to a minimum value of 37.9 W and is therefore three times lower than for conventional setups using circular coils [6]. These excellent results state previous simulation studies [8] and make FFL imaging to a promising alternative compared to FFP imaging.

Furthermore, the field generating parts were embedded in a customized scanner case that includes an air cooling device. This provides the basis for future work. The constructed field generator shall be extended to a full scanner setup in near future. Besides the possibility to generate an FFL, the scanner has the capacity of generating an FFP as well. The combination of both spatial encoding schemes in one scanner allows a fair practical comparison and gives information of the real sensitivity gain. Then, published simulations studies can be investigated and comprehended.

A further aim for this scanner setup is real-time in-vivo imaging of mice. The worlds first in-vivo mice images generated by a dynamic FFL configuration shall be realized in the future. Considering that, further steps have to be carried out. Transmission and receive chains are currently built up and have to be integrated in the system. Furthermore, amplifiers, i/o cards, and controlling are needed. Here, extended analyses between system matrix and Radon-based reconstruction techniques in a practical setup proof simulation studies from this area [7].

The implementation of curved rectangular coils building the basis of this work is not just feasible, but also offers new ideas for future coil optimizations. Used coil forms, manufacturing techniques and coil configuration methods can also be applied to future scanner types. Prospective FFP scanner designs could benefit from the results achieved in this thesis. Hence, higher gradient strengths can be achieved improving the resolution at moderate cooling effort.

[7] T. Knopp, M. Erbe, T. Sattel, S. Biederer, and T. Buzug. *A Fourier slice theorem for magnetic particle imaging using a field-free line*. Inverse Probl, 27(9):095004, 2011.

[8] M. Erbe, T. F. Sattel, T. Knopp, and T. M. Buzug. "Enhancing the Efficiency of a Field Free Line Scanning Device for Magnetic Particle Imaging". *IEEE Medical Imaging Conference, Anaheim, USA*, 2012.

[9] W. Nolting. *Grundkurs Theoretische Physik 3: Elektrodynamik*. Springer, Berlin, 2001.

[10] W. Demtröder. *Experimentalphysik 2: Elektrizität und Optik*. Springer, Berlin, 2004.

[11] T Knopp, S Biederer, T Sattel, J Weizenecker, B Gleich, J Borgert, and T. M. Buzug. *Trajectory analysis for magnetic particle imaging*. Phys Med Biol, 54(2):385–397, 2009.

[12] G. Lehner. *Elektromagnetische Feldtheorie: für Ingenieure und Physiker*. Springer, Berlin, 2008.

Chapter 7

Bibliography

7 Bibliography

[1] B. GLEICH and J. WEIZENECKER. *Tomographic imaging using the nonlinear response of magnetic particles.* Nature, 435(7046):1214–1217, 2005.

[2] J. WEIZENECKER, B. GLEICH, J. RAHMER, H. DAHNKE, and J. BORGERT. *Three-dimensional real-time in vivo magnetic particle imaging.* Phys Med Biol, 54(5):1–10, 2009.

[3] P. W. GOODWILL, E. U. SARITAS, L. R. CROFT, T. N. KIM, K. M. KRISHNAN, D. V. SCHAFFER, and S. M. CONOLLY. *X-Space MPI: Magnetic Nanoparticles for Safe Medical Imaging.* Adv Mater, 24(28):3870–3877, 2012.

[4] J. WEIZENECKER, B. GLEICH, and J. BORGERT. *Magnetic particle imaging using a field free line.* J Appl Phys, 41(10):105009, 2008.

[5] T. KNOPP, M. ERBE, S. BIEDERER, T. F. SATTEL, and T. M. BUZUG. *Efficient generation of a magnetic field-free line.* Med Phys, 37(7):3538–3540, 2010.

[6] M. ERBE, T. KNOPP, T. F. SATTEL, S. BIEDERER, and T. M. BUZUG. *Experimental generation of an arbitrarily rotated field-free line for the use in magnetic particle imaging.* Med Phys, 38(9):5200–5207, 2011.

[7] T. KNOPP, M. ERBE, T. F. SATTEL, S. BIEDERER, and T. M. BUZUG. *A Fourier slice theorem for magnetic particle imaging using a field-free line.* Inverse Probl, 27(9):095004, 2011.

[8] M. ERBE, T. F. SATTEL, T. KNOPP, and T. M. BUZUG. "Enhancing the Efficiency of a Field Free Line Scanning Device for Magnetic Particle Imaging". *IEEE Medical Imaging Conference, Anaheim, USA,* 2012.

[9] W. NOLTING. *Grundkurs Theoretische Physik 3: Elektrodynamik.* Springer, Berlin, 2001.

[10] W. DEMTRÖDER. *Experimentalphysik 2: Elektrizität und Optik.* Springer, Berlin, 2004.

[11] T. KNOPP, S. BIEDERER, T. F. SATTEL, J. WEIZENECKER, B. GLEICH, J. BORGERT, and T. M. BUZUG. *Trajectory analysis for magnetic particle imaging.* Phys Med Biol, 54(2):385–397, 2009.

[12] G. LEHNER. *Elektromagnetische Feldtheorie: für Ingenieure und Physiker.* Springer, Berlin, 2008.

[13] S. CHIKAZUMI. *Physics of Ferromagnetism (International Series of Monographs on Physics).* Oxford University Press, Oxford, 1997.

[14] J. RAHMER, J. WEIZENECKER, B. GLEICH, and J. BORGERT. *Signal encoding in magnetic particle imaging: properties of the system function.* BMC Med Imaging, 9:4, 2009.

[15] C. P. BEAN and J. D. LIVINGSTON. *Superparamagnetism.* J Appl Phys, 30(4):120, 1959.

[16] T. KNOPP. *Effiziente Rekonstruktion und alternative Spulentopologien für Magnetic-Particle-Imaging.* Vieweg+Teubner Verlag, Wiesbaden, 2011.

[17] R. LAWACZECK, H. BAUER, T. FRENZEL, M. HASEGAWA, Y. ITO, K. KITO, N. MIWA, H. TSUTSUI, H. VOGLER, and H. J. WEINMANN. *Magnetic iron oxide particles coated with carboxydextran for parenteral administration and liver contrasting. Pre-clinical profile of SH U555A.* Acta Radiol, 38(4 Pt 1):584–597, 1997.

[18] R. M. FERGUSON, K. R. MINARD, A. P. KHANDHAR, and K. M. KRISHNAN. *Optimizing magnetite nanoparticles for mass sensitivity in magnetic particle imaging.* Med Phys, 38(3):1619–1626, 2011.

[19] J. C. MAXWELL. *On physical lines of force.* Philosophical Magazine, 21(4), 1861.

[20] J. C. MAXWELL. *A Dynamical Theory of the Electromagnetic Field.* Phil Trans R Soc Lond, 155:459–512, 1865.

[21] J. C. MAXWELL. *Treatise on Electricity and Magnetism, Vol. 1.* Dover Publications, New York, 1954.

[22] D. C. GIANCOLI. *Physik: Lehr- und Übungsbuch.* Pearson Studium, München, 2010.

[23] T. M. BUZUG. *Computed Tomography: From Photon Statistics to Modern Cone-Beam CT.* Springer, Berlin, 2008.

[24] W. A. KALENDER. *Computed Tomography.* John Wiley & Sons, New Yersey, 2011.

[25] M. RIEGER, H. SPARR, R. ESTERHAMMER, C. FINK, R. BALE, B. CZERMAK, and W. JASCHKE. *Modern CT diagnosis of acute thoracic and abdominal trauma.* Anaesthesist, 51(10):835–842, 2002.

[26] Z.-P. LIANG and P. C. LAUTERBUR. *Principles of Magnetic Resonance Imaging: A Signal Processing Perspective.* Wiley-IEEE Press, New Jersey, 1999.

[27] E. M. HAACKE, R. W. BROWN, M. R. THOMPSON, and R. VENKATESAN. *Magnetic Resonance Imaging: Physical Principles and Sequence Design.* Wiley-Liss, New York, 1999.

[28] K. S. NAYAK and B. S. HU. *Triggered real-time MRI and cardiac applications.* Magn Reson Med, 49(1):188–192, 2003.

[29] D. BAILEY, D. TOWNSEND, P. VALK, and M. MAISEY. *Positron Emission Tomography: Basic Sciences.* Springer, Berlin, 2005.

[30] M. N. WERNICK and J. N. AARSVOLD. *Emission Tomography: The Fundamentals of PET and SPECT.* Academic Press, London, 2004.

[31] R. ENGLISH and S. BROWN. *Spect Single-Photon Emission Computed Tomography: A Primer.* Society of Nuclear Medicine, New York, 1986.

[32] K. BAUMANN, B. RUHLAND, K. HEINRICH, K. LÜDTKE-BUZUG, T. M. BUZUG, K. DIEDRICH AND D. FINAS. "Magnetic Particle Imaging durch superparamagnetische Nanopartikel zur Sentinellymphknotendetektion beim Mammakarzinom". *Senologie – Zeitschrift für Mammadiagnostik und -therapie,* 2011.

[33] U. VERONESI, G. PAGANELLI, G. VIALE, A. LUINI, S. ZURRIDA, V. GALIMBERTI, M. INTRA, P. VERONESI, C. ROBERTSON, P. MAISONNEUVE, G. RENNE, C. DE CICCO, F. DE LUCIA, and R. GENNARI. *A randomized comparison of sentinel-node biopsy with routine axillary dissection in breast cancer.* N Engl J Med, 349(6):546–553, 2003.

[34] R. W. KATZBERG and C. HALLER. *Contrast-induced nephrotoxicity: clinical landscape.* Kidney Int Suppl, (100):3–7, 2006.

7 Bibliography

[35] P. A. MCCULLOUGH. *Contrast-induced acute kidney injury.* J Am Coll Cardiol, 51(15):1419–1428, 2008.

[36] P. H. KUO, E. KANAL, A. K. ABU-ALFA, and S. E. COWPER. *Gadolinium-based MR contrast agents and nephrogenic systemic fibrosis.* Radiology, 242(3):647–649, 2007.

[37] J. HAEGELE, J. RAHMER, B. GLEICH, C. BONTUS, J. BORGERT, H. WOJTCZYK, T. M. BUZUG, J. BARKHAUSEN, and F. M. VOGT. *Visualization of Instruments for Cardiovascular Intervention Using MPI. Magnetic Particle Imaging: A Novel Spio Nanoparticle Imaging Technique,* 140:211, 2012.

[38] H. WOJTCZYK, J. HÄGELE, M. GRÜTTNER, W. TENNER, G. BRINGOUT, M. GRÄSER, F. M. VOGT, J. BARKHAUSEN, and T. M. BUZUG. *Visualization of Instruments in interventional Magnetic Particle Imaging (iMPI): A Simulation Study on SPIO Labelings.* Springer Proc Phys, 140:167–172, 2012.

[39] P. W. GOODWILL and S. M. CONOLLY. *The X-space formulation of the magnetic particle imaging process: 1-D signal, resolution, bandwidth, SNR, SAR, and magnetostimulation.* IEEE Trans Med Imaging, 29(11):1851–1859, 2010.

[40] A. KRAMLICH, J. BOHNERT, and O. DÖSSEL. *Transmembrane Voltages Caused by Magnetic Fields – Numerical Study of Schematic Cell Models.* Springer Proc Phys, 140:337–342, 2012.

[41] G. BRINGOUT, H. WOJTCZYK, M. GRÜTTNER, M. GRÄSER, W. TENNER, J. HÄGELE, F. M. VOGT, J. BARKHAUSEN, and T. M. BUZUG. *Safety Aspects for a Pre-clinical Magnetic Particle Imaging Scanner.* Springer Proc Phys, 140:355–359, 2012.

[42] E. U. SARITAS and P. W. GOODWILL. *Safety Limits for Human-Size Magnetic Particle Imaging Systems.* Springer Proc Phys, 140:325–330, 2012.

[43] P. W. GOODWILL and S. M. CONOLLY. *Multidimensional x-space magnetic particle imaging.* IEEE Trans Med Imaging, 30(9):1581–1590, 2011.

[44] P. W. GOODWILL, J. J. KONKLE, B. ZHENG, E. U. SARITAS, and S. M. CONOLLY. *Projection x-space magnetic particle imaging.* IEEE Trans Med Imaging, 31(5):1076–1085, 2012.

[45] P. W. GOODWILL and J. J. KONKLE. *Projection X-space MPI Mouse Scanner.* Springer Proc Phys, 140:267–271, 2012.

[46] P. W. GOODWILL and L. R. CROFT. *Third Generation X-space MPI Mouse and Rat Scanner.* Springer Proc Phys, 140:261–265, 2012.

[47] T. KNOPP, J. RAHMER, T. F. SATTEL, S. BIEDERER, J. WEIZENECKER, B. GLEICH, J. BORGERT, and T. M. BUZUG. *Weighted iterative reconstruction for magnetic particle imaging.* Phys Med Biol, 55(6):1577–1589, 2010.

[48] J. WEIZENECKER, J. BORGERT, and B. GLEICH. *A simulation study on the resolution and sensitivity of magnetic particle imaging.* Phys Med Biol, 52(21):6363–6374, 2007.

[49] B. GLEICH, J. WEIZENECKER, and J. BORGERT. *Experimental results on fast 2D-encoded magnetic particle imaging.* Phys Med Biol, 53(6):81–84, 2008.

[50] T. KNOPP, T. F. SATTEL, S. BIEDERER, J. RAHMER, J. WEIZENECKER, B. GLEICH, J. BORGERT, and T. M. BUZUG. *Model-based reconstruction for magnetic particle imaging.* IEEE Trans Med Imaging, 29(1):12–18, 2010.

[51] T. KNOPP, S. BIEDERER, T. F. SATTEL, J. RAHMER, J. WEIZENECKER, B. GLEICH, J. BORGERT, and T. M. BUZUG. *2D model-based reconstruction for magnetic particle imaging.* Med Phys, 37(2):485–491, 2010.

[52] J. RADON. *Über die Bestimmung von Funktionen durch ihre Integralwerte längs gewisser Mannigfaltigkeiten.* Ber Verh Sächs Akad Wiss, 69:262–277, 1917.

[53] R. C. GONZALEZ and R. E. WOODS. *Digital Image Processing.* Prentice Hall, New Jersey, 2002.

[54] J. S. LIM. *Two-Dimensional Signal and Image Processing.* Prentice Hall PTR, New Jersey, 1989.

[55] T. KNOPP, T. F. SATTEL, S. BIEDERER, and T. M. BUZUG. *Field-Free line formation in a magnetic field.* J Phys A, 43(1):012002, 2010.

[56] E. C. JORDAN and K. G. BALMAIN. *Electromagnetic Waves and Radiating Systems.* Prentice Hall, New Jersey, 1968.

[57] J. F. KENNEY and E. S. KEEPING. *Mathematics of Statistics: Pt. 1.* Princeton, New Jersey, 1954.

Wir verlegen wissenschaftlichen Schriften

Bachelor- und Masterarbeiten,
Dissertationen und Habilitationen,
Monografien und Tagungsbände, etc.

Kostenlose Verlegung als Buch mit ISBN-Nummer
und Aufnahme in die Deutsche
Nationalbibliothek

Hochwertiger Buchdruck in nachhaltiger
Produktion (FSC-zertifiziert)

Günstiger Bezug von Autorenexemplaren
Weltweite Präsenz des Werkes bei den
großen Händlern: Amazon, Thalia,
Hugendubel, Barnes & Noble u.v.m. sowie
optional als eBook

www.infinite-science.de/publishing

Infinite Science GmbH
MFC 1 | BioMedTec Wissenschaftscampus
Maria-Goeppert-Str. 1, 23562 Lübeck
book@infinite-science.de